SPRINGER
LAB MANUAL

Springer

Berlin
Heidelberg
New York
Barcelona
Budapest
Hong Kong
London
Milan
Paris
Santa Clara
Singapore
Tokyo

Johannes Schenkel (Ed.)

RNP Particles, Splicing and Autoimmune Diseases

With 49 Figures

Springer

DR. JOHANNES SCHENKEL
II. Physiologisches Institut
der Universität Heidelberg
Im Neuenheimer Feld 326
D-69120 Heidelberg

ISBN 978-3-642-48975-4   ISBN 978-3-642-80356-7 (eBook)
DOI 10.1007/978-3-642-80356-7

Library of Congress Cataloging-in-Publication Data
RNP particles, splicing, and autoimmune diseases / Johannes Schenkel, ed. p. cm. -- (Springer lab
manual) Includes bibliographical references and index. ISBN (invalid) 3-540-62448-7 (wire-o-
binding : alk. paper) 1. Ribosomes--Research--Laboratory manuals. 2. RNA splicing--Laboratory
manuals. 3. Nucleoproteins--Research--Laboratory manuals. 4.Autoimmune diseases--Research--
Laboratory manuals. I. Schenkel, Johannes. II. Series. [DNLM: 1. Ribonucleoproteins--laboratory
manuals. 2. RNA Splicing--laboratory manuals. 3. Autoantibodies--chemistry--laboratory manuals. 4. Autoimmune Diseases--diagnosis--laboratory manuals. QU 25 R627 1998] QH603. R5R65
1998 572.8'8--dc21 DNLM/DLC for Library of Congress

Production: PRO EDIT GmbH, D-69126 Heidelberg
Cover design: design & production GmbH, D-69121 Heidelberg
Typesetting: Mitterweger Werksatz GmbH, D-68723 Plankstadt
SPIN 10523165   31/3137   5 4 3 2 1 0-Printed on acid free paper

Preface

The first insights into the site and mechanisms of RNA processing to functional mRNA in eukaryotic cells came from the group of Georgiev (Lukanidin et al. 1972) who demonstrated the association of rapidly labelled, heterogeneous nuclear RNA (hnRNA) with a limited number of specific proteins in the cell nucleus. These "informofers", i.e. packaged precursors of mRNA (pre-mRNA or hnRNA), are in a form presumably amenable to the action of nucleases. With the availability of better analytical techniques, the considerable heterogeneity of hnRNA associated proteins was revealed (Niessing and Sekeris 1970), suggesting a role that was more composite, rather than solely structural, for these proteins. Later studies investigated the RNA binding behavior of these proteins (Schenkel et al. 1988, 1989; Wilk et al. 1983).

For a long time, the small nuclear RNAs, well characterized with respect to primary structure (reviewed by Reddy and Busch 1983), were naively ignored regarding their function. Several events then set the stage for a detailed study of the intricate mechanisms of the splicing process and other steps involved in hnRNA processing: (1) The demonstration of a second class of nuclear ribonucleoproteins (RNPs), composed of small nuclear RNAs (snRNAs) and another characteristic group of heterogeneous proteins (Lerner et al. 1980; Guialis et al. 1983); (2) the detection of the association of snRNPs with hnRNPs by virtue of base pairing between hnRNA and snRNA (Flytzanis et al. 1978); (3) the possible role of hnRNA/snRNA hybrids at exon/intron boundaries as sites of nuclease cleavage (Mount et al. 1983); (4) more detailed studies on spliced genes and RNA splicing (among others: Krämer and Keller 1990; Kastner at al. 1992; Mehlin et al. 1992; Izaurralde and Mattaj 1995).

The realization that in many autoimmune diseases many of the autoantibodies are directed against snRNP components allowed both the isolation and characterization of snRNPs (Padgett et al. 1983, van Dam et al. 1989) and the diagnosis, prediction and treatment of autoimmune diseases.

An introduction to the vast field of nuclear RNP particles and their role in splicing and in autoimmune diseases, both from the basic research and the clinical viewpoint, was the object of a FEBS practical and lecture course held in the fall of 1995 in Athens. An interdisciplinary approach to this field was understandably necessary, with participation of biochemists, molecular and cell biologists and clinicians. Both the structural aspects of nuclear RNPs and their functional significance were stressed. Methods of isolation and characterization of RNPs, including immunochemical techniques, were applied. Splicing experiments were also performed using suitable nuclear extracts and model genes and their transcription products. The course also served to acquaint participants with other current biochemical, clinical chemical and molecular biological methods.

Selected specific techniques which were used or discussed in the course have been included in this manual by the lecturers, who are acknowledged specialists in their fields and who routinely use these techniques. The chapters in this manual are:

Isolation and Immunochemical Characterization of hnRNP Particles; The lnRNP Particle – A Naturally Assembled Complex of Pre-mRNA and Splicing Factors; Intrinsic Fluorescence Techniques for Studies on Protein-Protein and Protein-RNA Interactions in RNP Particles; Procedures for Three-Dimensional Reconstitution from Thin Sections with Electron Tomography; Purification and Electron Microscopy of Spliceosomal snRNPs; Detection of Autoantibodies to Extractable Cellular Antigens; Methods in Immunolocalization of Autoantigens; In Vitro Splicing of Pre-mRNA.

Heidelberg, October 1997   JOHANNES SCHENKEL

References

Flytzanis C, Alonso A, Louis C, Krieg L, Sekeris CE (1978) Association of small nuclear RNA with hnRNA isolated from nuclear RNP complexes carrying hnRNA. FEBS Lett 96: 201–206

Guialis A, Arvanitopoulou A, Patrinou-Georgoula M, Sekeris CE (1983) Identification of two discrete ribonucleoprotein particles within the monomer population of rat liver nuclear RNPs. FEBS Lett 151: 127–133

Izaurralde E, Mattaj IW (1995) RNA export. Cell 81: 153–159

Lerner MR, Boyle JA, Mount SM, Wolin SL, Steitz JA (1980) Are snRNPs involved in splicing? Nature 283: 220–224

Kastner B, Kornstadt U, Bach M, Lührmann R (1992) Structure of the small nuclear RNP particle U1: identification of the two structural protuberrances with RNP-antigens A and 70K. J Cell Biol 116: 839–849

Krämer A, Keller W (1990) Preparation and fractionation of mammalian extracts active in pre-mRNA splicing. In: Dahlberg JE, Abelson IN (eds) Methods in enzymology, vol 181. Academic Press, London, pp 3–19

Lukanidin EM, Zalmanzon ES, Komaromi L, Samarina OP, Georgiev GP (1972) Structure and function of informofers. Nature New Biol 285: 193–197

Mehlin H, Daneholt B, Skoglund U (1992) Translocation of a specific pre-messenger ribonucleoprotein particle through the nuclear pore studied with electron microscope tomography. Cell 69: 605–613

Mount SM, Petterson I, Hinterberger M, Karmas A, Steitz JA (1983) The U1 small nuclear RNA-protein complex selectively binds a 5' splice site in vitro. Cell 33: 509–518

Niessing J, Sekeris CE (1970) Cleavage of high-molecular weight DNA-like RNA by a nuclease present in 30-S ribonucleoprotein particles of rat liver nuclei. Biochim Biophys Acta 209: 484–492

Padgett RA, Mount SM, Steitz JA, Sharp P (1983) Splicing of messenger RNA precursors is inhibited by antisera to small nuclear ribonucleoprotein. Cell 35: 101–107

Reddy R, Busch H (1983) Small nuclear RNAs and RNA processing. Prog Nucleic Acid Res Mol Biol 30: 127–162

Schenkel J, Sekeris CE, Alonso A, Bautz EKF (1988) RNA-binding properties of hnRNP proteins. Eur J Biochem 171: 565–569

Schenkel J, Appel I, Schwarzwald R, Bautz EKF, Wolfrum J, Greulich KO (1989) Fluorescence studies on the role of tryptophan in hnRNP particles of HeLa cells. Biochem J 263: 379–283

van Dam A, Winkel I, Zijlstra-Baalbergen J, Smeenk R, Cuypers HT (1989) Cloned human snRNP proteins B and B' differ only in their carboxy-terminal part. EMBO J 8: 3853–3860

Wilk HE, Angeli G, Schäfer KP (1983) In vitro reconstitution of 35S ribonucleoprotein complexes. Biochemistry 22(19): 4592–4600

Contents

Chapter 1

Isolation and Immunochemical Characterization of hnRNP Particles

Frank Jung[1], Constantin E. Sekeris[2],
and Johannes Schenkel[3*]

Introduction

The processing of messenger RNA precursors (pre-messenger RNA or pre-mRNA, heterogeneous nuclear RNA or hnRNA) to mature mRNA molecules and its transport from the nucleus to the cytoplasm is a multistep process: A modified nucleotide (cap structure) is added to the 5' end of the primary transcript, intervening sequences (introns) are spliced and the 3' ends are poly-adenylated.

hnRNA is packaged by proteins forming protein/RNA complexes, which are of special interest, since they are the site of the posttranscriptional modifications of RNA. The RNA in the complexes is associated with the proteins in a similar manner like DNA with histones during nucleosome formation. RNP particles can be organized as monomers, oligomers or polymers. If the RNA joining the RNPs is cleaved by nucleases or mechanically during preparation, monomeric hnRNP complexes are isolated sedimenting at approximatively 40S. These monomeric particles consist of RNA and several proteins. The best described are the highly conserved core proteins A_1, A_2, B_1, B_2, C_1 and C_2, ranging between 32 and 45 kDa. This field has been recently reviewed (Biamonti and Riva 1994; Burd and Dreyfuss 1994; Dreyfuss et al. 1993; Görlach et al. 1993; Izaurralde and Mattaj 1995; Mattaj and Nagai 1995; Nagai et al. 1995; Nigg et al. 1991; Schenkel et al. 1988; Swanson 1990).

---

* Corresponding author: Johannes Schenkel: Tel.: (+49)-6221–54–4064;
  Fax: (+49)-6221–54–6364; e-mail: dj5@ix.urz.uni-heidelberg.de
[1] Institute of Molecular Genetics, University of Heidelberg, Germany
[2] The National Hellenic Research Foundation, Institute of Biological
  Research and Biotechnology, Athens, Greece
[3] II. Institute of Physiology, University of Heidelberg,
  Im Neuenheimer Feld 326, 69120 Heidelberg, Germany

Other RNP particles, the snRNPs (small nuclear RNPs), are also involved in the processing of pre-mRNA. These complexes contain small nuclear RNA species (60–215 nucleotides). This class of RNA is very stable and contains high amounts of uridine. Frequently detected in snRNPs are the RNA species U1, U2, U4, U5 and U6. These particles can be isolated from nucleic fractions as 10S–20S particles, each containing one or two RNA species (U1, U2, U4/6 or U5) and snRNP-specific proteins. SnRNP, hnRNP and RNA/RNA interactions are responsible for spliceosome formation and subsequently for the splicing of pre-mRNA (Ares and Weiser 1995; Hodges and Beggs 1994; Lührmann 1990; Lührmann et al. 1990; Newman 1994a, Will et al.1993).

Several human antibodies were used to examine the components of RNP particles and to delineate their role in RNA processing. These antibodies were detected in sera of patients suffering from autoimmune diseases, such as sytemic lupus erythematosus (SLE). They also play an important role in the development of in vitro splicing and in vitro 3' end processing systems (Hodges and Bernstein 1994; Lamm and Lamond 1993; Lamond 1993; Medhani and Guthrie 1994; Newman 1993, 1994b; Nilsen 1994; Wittop et al. 1994).

1.1
Preparation of hnRNP Particles

A prerequisite to isolating RNP is the preparation of nuclei avoiding ribosomal contaminations. For many RNP experiments extracting RNP from tissue culture cells (e.g. HeLa cells) or from livers (e.g. livers of rats or mice) is recommended. To avoid coextraction of nucleosomes, very careful preparation of RNP from the nuclei is required (Louis and Sekeris 1976; Schenkel et al. 1988).

Materials

Motor-driven homogenizer with a Teflon pestle (Potter homogenizer); homogenizer with a glass pestle (Dounce homogenizer); Eppendorf centrifuge; refrigerated lab centrifuge; ultracentrifuge equipped with swinging bucket rotors; density gradient fractionator (optional); photometer (optional); refractometer (optional); UV-fluorescence microscope (optional); sonifier; gradient mixer; cheese cloth.

- TSS: 50 mM Tris-HCl, pH 7.6; 10 mM $MgCl_2$; 25 mM KCl
- TSS with sucrose and PMSF: 1 mM phenylmethyl sulfonylfluoride (PMSF, freshly prepared) and 250 mM sucrose (RNase-free) in TSS
- NHS: 20 mM Tris-HCl, pH 7.2; 1.5 mM $MgCl_2$; 0.02 % Triton X-100
- pH 8 Buffer: 10 mM Tris-HCl, pH 8; 140 mM NaCl; 1 mM $MgCl_2$
- Sucrose solutions: sucrose concentrations as recommended in TSS, e.g., 10 and 40 % sucrose solutions
- TSS/EDTA: 50 mM EDTA in TSS
- Ethidium bromide: 10 mg/ml ethidium bromide in distilled H_2O

Procedure

Preparation of Nuclei from HeLa S3 Cells

1. Collect 10^9 HeLa cells or thaw the same number of frozen cells rapidly.

2. Suspend cells in 20 ml TSS with sucrose and PMSF.

3. Prepare nuclei by homogenization in a motor-driven Potter homogenizer: 20 strokes, 1000 rpm, 4 °C.

4. Spin homogenate to separate cytolasm and nuclei 10 min, 2600 rpm, 4 °C. Use pellet to purify nuclei, use supernatant to isolate cytoplasmic components.

5. Remove polysomes from the nuclear membrane: Resuspend nuclei (pellet of step 4) in 50 ml NHS/10^9 cells, allow to stand 10 min at 4 °C. Treat nuclei in a motor-driven Potter homogenizer (7 strokes, 1000 rpm).

6. Spin to collect nuclei 10 min, 400 g, 4 °C.

7. Resuspend nuclei (pellet of step 6) in 50 ml NHS/10^9 cells. Homogenize and spin as described in steps 5 and 6.

8. To remove the bulk of detergent, resuspend nuclei (pellet of step 7) in 30 ml ice cold distilled H_2O. Resuspend using a 10 ml glass pipette. Work quickly! Collect nuclei by centrifugation 10 min, 650 g, 4 °C.

9. Take a small aliquot of the nuclei (pellet of step 8), stain with a drop of ethidium bromide and analyze the quality of preparation in an UV fluorescence microscope. The nuclei will shine brightly under UV-fluorescence, cytoplasmic contaminations weakly.

Preparation of Nuclei from Rat Livers

1. Prepare livers from ca. 10 rats (150 g animals).

2. Wash in TSS and weigh.

3. Mince with scissors into small pieces.

4. Homogenize in three volumes of TSS $+1$ mM PMSF in a Potter homogenizer (loose fit=L).

5. Filter homogenate through six layers of cheese cloth.

6. Spin 10 min at 2600 g, 4 °C. Use pellet to clarify nuclei, use supernatant to isolate cytoplasmic components.

7. Resuspend nuclei in 2.2 M sucrose in TSS (ca. 13.5 ml/liver).

8. Homogenize four times in a loose fit homogenizer.

9. Load nuclei (suspension of step 8) on an 8 ml 2.2 M sucrose in TSS cushion.

10. Spin 2 h in a Beckman SW 27 rotor 24 000 rpm, 4 °C.

11. Remove supernatant, clean wall of the tube with Kleenex.

12. Remove polysomes from the nuclear membrane: Resuspend nuclei (pellet of step 10) in 50 mM EDTA in TSS (ca. 8 ml/liver). Homogenize carefully in a loose fit homogenizer.

13. Spin nuclei 10 min 650 g, 4 °C.

14. Remove supernatant, resuspend nuclei (pellet of step 13) in pH 8 buffer (ca. 1 ml/liver).

15. Stain aliquot of nuclei (pellet of step 13) with a drop of ethidium bromide and analyze quality of preparation in the UV fluorescence microscope. Nuclei will shine brightly under the UV fluorescence, cytoplasmatic contaminations weakly.

Isolation of hnRNP Particles

1. Resuspend nuclei in 2 ml pH 8 buffer/10^9 cells or two livers. Allow nuclei to swell for 45 min. Resuspend gently repeated times.

2. Extract RNP by three sonications (10 s each with a 30 s break) at 4 °C. **Note:** Do not touch the tube with the sonifier!

3. Spin 10 min, 16 500 g, 4 °C to remove remnants of broken nuclei. Use hnRNP containing supernatant for further experiments.

4. To prepare 40S particles from cells with a low activity of endogenous RNases, incubate supernatant of step 3 for about 30 min at 37 °C to activate the endogenous RNases which cleave the RNA and generate 40S particles. This step is to be omitted in tissues with high endogenous RNase activity.

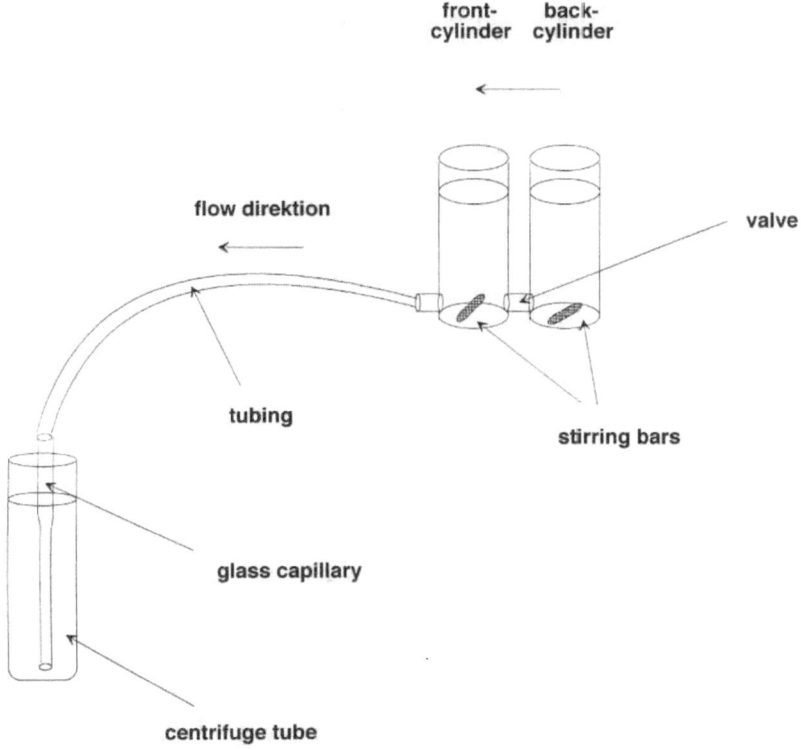

Fig. 1.1. A gradient mixer set to generate sucrose gradients

5. Prepare sucrose gradients: Put gradient mixer on a magnetic stirrer, making sure that the glass capillary touches the bottom of the centrifuge tube. Close valve. Fill the rear cylinder of the gradient mixer with the higher concentration sucrose solution. Lock tubing with a clip. Fill front cylinder with low concentration sucrose solution, put magnetic stirring bars into both cylinders, e.g., 10 and 40 %. Start magnetic stirrer, remove clip from the tubing. When low concentration sucrose solution starts running, open valve. If this does not happen, press air into the front cylinder with your hand or finger (use glove!) before unlocking the valve (see Fig. 1.1). **Note:** Make sure that all gradients prepared in parallel have the same volume!

6. Load RNP containing supernatant on sucrose gradients (for example 10 %–40 % sucrose gradients in TSS). Spin at 4 °C until 40S peak is in the middle of the gradient. Calculate conditions of spin as described in the next chapter.

7. Fractionate gradient in about 20 fractions using a density gradient fractionator. Detect peaks by monitoring the OD$_{260}$. **Note:** In case of failure of the fractionator take fractions carefully from the top of the gradient using a pipette. Monitor the peaks of one gradient only.

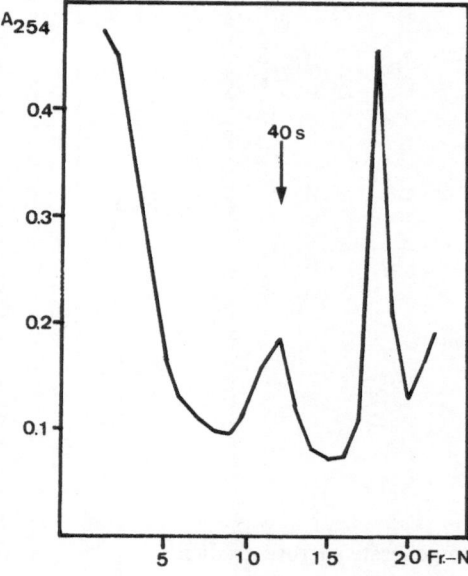

Fig. 1.2. Sedimentation profile: monomeric hnRNP particles isolated from HeLa cells separated in a 10–40 % sucrose gradient using a Beckman SW 40 rotor, $\omega^2 t = 5.4 \times 10^{11}$

8. Get RNP peak fractions. If concentration is needed, RNP can be precipitated by ethanol or by sedimentation in an ultra-centrifuge.
   An OD_{260} profile of hnRNP separated on a sucrose gradient is shown in Fig. 1.2.

▨ Troubleshooting

● The recovery of hnRNP protein after fractionation is too low.
  – The amount of cells used was insufficient or sonication was not efficient. Start the preparation again with a sufficient amount of cells. Increase energy of sonification.

● Nucleosomal or histone contamination in hnRNP preparation.
  – Coextraction of nucleosomes with hnRNP. Check pH of buffers used during extraction; reduce energy applied by sonifier.

▨ Results

Calculation of S Values in the Sucrose Gradients

The distance in the gradient that has been covered by the material to be separated by centrifugation depends on several factors:

– The concentration of sucrose in the gradient
– The rotor and its diameter
– The speed of the rotor
– The temperature
– The duration of the run
– The density of the particles

To calculate the sedimentation values, properties of the rotor and the gradient are taken into consideration by the equation:

$$z_0 = (z_1 \times r_2 - z_2 \times r_1) : r_2 - r_1$$

where r_1 is the distance from the rotor axis to the top of the gradient; r_2 is the distance from the rotor axis to the bottom of the gradient, z_1 is the concentration of sucrose at the top of the gradient; z_2 is the concentration of sucrose at the bottom of the gradient. The corresponding integrals for single sucrose concentrations are described by McEwen (1967).

Calculate S values by the equation:

$$S = (I_{(zx)} - I_{(z1)}) : \omega^2 \times t \times 10^{-13}$$

where $I_{(zx)}$ is the time integral of the sucrose concentration in the corresponding fraction; $I_{(z1)}$ is the time integral of the sucrose concentration at the top of the gradient; ω^2 is equal to $(0.10472 \times rpm)^2$; t is the time of centrifugation in seconds.

Determine sucrose concentration by refractometry or by a calibration curve.

1.2
Gel Electrophoresis of Protein

One of the most important methods for the characterization of proteins is SDS-polyacrylamide gel electrophoresis (SDS-PAGE). The proteins are denatured by the detergent sodium dodecyl-sulfate (SDS); possible disulfide links in the proteins or in dimers are cleaved by β-mercaptoethanol (β-ME). Thus, the separation of the proteins depends only on the molecular range of the protein and not on secondary or tertiary structures or the charge of the protein (Laemmli 1970).

Depending on the pH and the ionic strength in the gel system, the proteins are concentrated in a stacking gel to a sharp band during the first part of the gel run. During the second part the proteins are separated depending on their molecular size (smaller proteins run a longer distance than larger proteins). The concentration of acrylamide to be used depends on the molecular range of the proteins to be analyzed. It is also possible to prepare gradient gels resulting in a wider range of separation.

Materials

- SDS sample solution: 10 % glycerol; 5 % β-mercaptoethanol; 3 % SDS; 10 mM Tris-HCl, pH 6.8; 0.02 % bromophenol blue
- 30 % acrylamide: 30 % acrylamide; 0.8 % N,N'methylene-bisacrylamide
- Upper Tris (stock solution): 6.06 g Tris-HCl, pH 6.8; 4.0 ml 10 % SDS. Final volume 100 ml with double-distilled H_2O; filter!

- Lower Tris (stock solution): 36.34 g Tris-HCl, pH 8.8; 8.0 ml 10 % SDS. Final volume 200 ml with double-distilled H_2O; filter!
- APS: 10 % ammonium persulfate in distilled H_2O
- TEMED (Tetramethylendiamine)
- Running buffer: 25 mM Tris-OH; 200 mM glycine; 0.1 % SDS
- Ethanol: 96 % ethanol and 80 % ethanol
- TCA: 100 % trichloroacetic acid in distilled H_2O
- Acetone

1. Add TCA to the protein samples to a final concentration of 10 %.

2. Precipitate proteins 60 min on ice.

3. Spin 10 min full speed in an Eppendorf centrifuge. Remove supernatant and wash twice with 80 % ethanol or acetone ($-20\,°C$).

4. Spin again and dry the pellet in a desiccator or vacuum centrifuge.

5. Alternatively, add two volumes of 96 % ethanol to the sample and precipitate RNP overnight at $-20\,°C$.

6. Spin 20 min full speed in an Eppendorf centrifuge. Dry pellet as described in step 4.

Preparation of Samples

1. Dissolve sample in sample buffer, making sure that the pH is alkaline (blue colour)! In case of a large volume, concentrate samples as described above.

2. Incubate samples for 2 min at 95 °C; spin afterwards in an Eppendorf centrifuge at full speed for 10 min to remove unsoluble material. If the color of the SDS-loading buffer becomes green or yellow, add NH_3 vapour until the colour changes to blue.

3. Load samples without insoluble material onto the slots of the gel.

Preparation of Gel

1. Rinse glass plates with alcohol.

2. Put spacers between the two glass plates, as shown in Fig. 7.4.

3. To seal, fill with boiled agarose around the spacers.

4. Estimate volume of gel.

5. Prepare separating gel by mixing 10 % gel solution (standard gel):
   - 10 ml 30 % acrylamide solution
   - 7.5 ml lower Tris
   - 12.5 ml distilled H_2O
   - 30 µl TEMED
   - 150 µl APS (10 %)

6. Immediately after mixing, pour gel solution between the glass plates (avoid air bubbles!); this is best done using a 10 ml glass pipette. Stop when the gel solution reaches ca. 3 cm short of the upper end of the shorter glass plate. **Note:** Depending on the size of the proteins to be separated, other concentrations of acrylamide may be recommended.

7. Alternatively, prepare gradient gel: **Refrigerated** gel solutions are placed in the gradient mixer, with the end of the tubing on the upper end of the gel. Rinse gradient mixer with water immediately after use. Estimate volume needed (Fig. 1.3).

8. Mix gradient gel solutions (for a 24 ml gel):
   - **Front:** 15 % gel solution:
   - 3 ml distilled H_2O
   - 3 ml lower Tris
   - 6 ml acrylamide solution
   - 15 µl TEMED
   - 22 µl 10 APS (10 %)
   - **Back:** 8 % gel solution:
   - 5.8 ml distilled H_2O
   - 3 ml lower Tris
   - 3.2 ml acrylamide solution
   - 15 µl TEMED
   - 20 µl 10 % APS (10 %)
   Use gradient mixer as described above (Fig. 1.3).

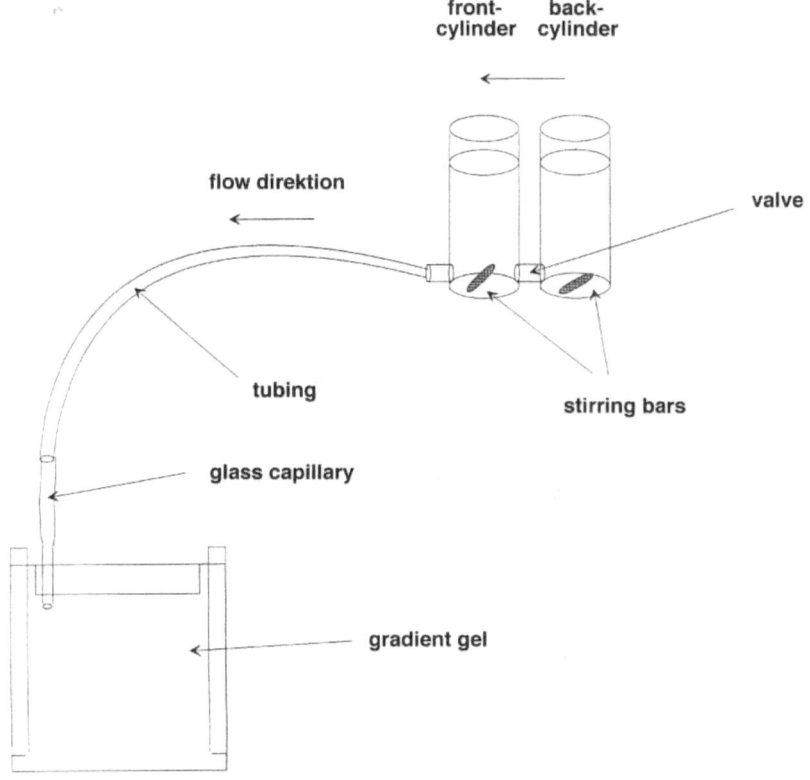

Fig. 1.3. A gradient mixer set to pour a polyacrylamide gel

9. Overlay the gel solution immediately and carefully with distilled water. Allow the gel to polymerize. The polymerization is completed if a sharp band (the surface of the gel) becomes visible.

10. Rinse the surface of the gel after polymerization with water and remove all remaining liquid with a paper towel.

11. Mix stacking gel solution:
    - 1.1 ml 30 % acrylamide solution
    - 2.5 ml upper Tris
    - 6.5 ml distilled H_2O
    - 20 μl TEMED
    - 40 μl 10 % APS

12. Overlay the separating gel with the stacking gel solution immediately after mixing. Put comb into the gel, avoiding air bubbles. Allow gel to polymerize. Use a few milliliters of gel solution as a polymerization marker.

13. Wet the surface of the gel with running buffer and remove the comb carefully. Rinse slots with running buffer and make sure that the slots are properly shaped.

14. Remove lower spacer and place gel on the gel apparatus. Fill the upper reservoir with running buffer and make sure that the system is not leaky. Fill lower reservoir with running buffer and remove air bubbles between the glass plates with a syringe.

Electrophoresis

1. Load samples (only soluble material) on the slots; avoid overloading.

2. Start electrophoresis at 100 V. When the bromophenol blue front reaches the border between the stacking and separating gels, allow the gel to run at 200 V.

3. For the second dimension of a two-dimensional gel electrophoresis, place the round gel on a stacking gel which has been filled between the 2-D glass plates. Seal round gel with 1 % agarose in O-buffer (see Sect. 1.4) onto the stacking gel.

4. When the bromophenol blue front reaches the lower end of the gel, interrupt power and remove glass plates with the gel from apparatus. Remove one glass plate by carefully lifting with a spatula. Allow gel to lie on the other glass plate and remove the stacking gel with a paper towel. Mark one corner of the gel and transfer the gel carefully onto a dish for staining.

1.3
Staining of Protein Gels

Materials

- Staining solution: 0.6 % w/v Coomassie brillant blue R-250; 40 % v/v methanol; 10 % v/v acetic acid in distilled H_2O; filter.
- Destaining solution: 20 % v/v methanol, 10 % v/v acetic acid in distilled H_2O
- TCA: 10 % solution in distilled H_2O
- Methanol: acetic acid: distilled H_2O solution (5:1.2:3.8)
- Fixation buffer: 1 % glutaraldehyde; 2 % $Na_2B_4O_7$ in distilled H_2O
- Silver nitrate solution: 20 mM NaOH; 4 % NH_4OH. Carefully add $AgNO_3$ to a final concentration of 20 %.
- Reducer: 10 % ethanol; 0.006 % citric acid; 0.01 % formaldehyde
- Background reducer: 5 % $Na_2S_2O_3$; 0.1 % $K_3Fe(CN)_6$ in distilled H_2O

Procedure

Coomassie Blue Stain

1. Use freshly prepared staining solutions only. Do not allow the gel to stain for longer than 20 min.

2. Destain the gel until the background is clear. Change the destaining solution a few times. The addition of small foam pads to the destaining solution helps to save destainer.

3. Recycle destaining solution: Filter the used solution through activated charcoal (loaded onto a paper filter). The filtrate can be used again as destainer.

Silver Stain

1. Fix proteins in the gel using 10 % TCA for 10 min.

2. Rinse gel twice with methanol:acetic acid:H_2O solution, 30 min.

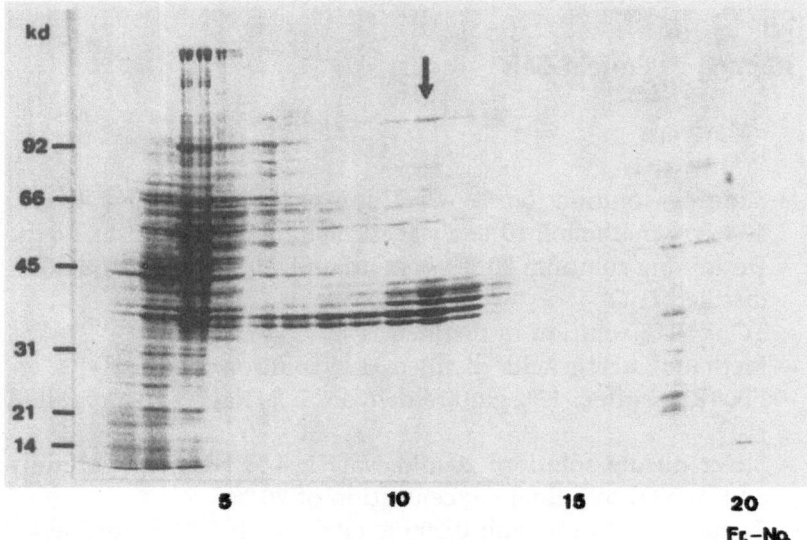

Fig. 1.4. Coomassie blue-stained 8–15% SDS gel of the RNP fractions shown in Fig. 1.2. *Arrow* indicates the 40S peak

3. Rinse gel three times with distilled H_2O. **Note:** For a Coomassie blue prestained gel, omit steps 1–3!

4. Fix gel 15 min in glutaraldehyde.

5. Rinse gel for 1 h in distilled H_2O, changing H_2O twice.

6. Stain proteins 15 min with silver nitrate solution.

7. Reduce silver nitrate until protein bands are visible. Stop reaction immediately with distilled H_2O.

8. Reduce possible background staining carefully by incubation in background reducer. Stop reaction by washing with distilled H_2O.

A Coomassie blue-stained protein gel of a hnRNP preparation, as shown in Fig. 1.2, is demonstrated in Fig. 1.4.

1.4
Isoelectric Focusing, Nonequilibrium pH Gradient Electrophoresis

In addition to separation of proteins on the basis of their molecular range they can also be separated on the basis of their charge isoelectrical point (IEP) by pH gradient electrophoresis. In two-dimensional separation, the sample is first submitted to isoelectric focusing (IEF) and then to SDS-PAGE. Cytochrome c is a useful marker protein for the first dimension (O'Farrell et al. 1977).

Nonequilibrium pH gradient electrophoresis (NEpHGE) is a relatively quick method to find out whether a protein is basic or acidic. NEpHGE cannot be used to detect the exact pH value of the isoelectric point of a protein. This can be done with IEF, but it is much more costly than NEpHGE.

Both gels are prepared as round gels in glass tubes. The pH gradient develops in an electrical field depending on the ampholytes and the chamber buffer used. After the run, the round gel is pushed out of the glass tube with a water-filled syringe and is afterwards equilibrated in a denaturing buffer (O-buffer). The gel is then layered on a specially prepared SDS gel. SDS gel electrophoresis as the second dimension of separation is as described above.

▨ Materials

- Lysis buffer I: 0.5 % SDS; 9.5 M urea; 5 % β-mercaptoethanol; 0.2 % ampholytes. **Note:** β-mercaptoethanol is a poison and very volatile.
- Lysis buffer II: 9.5 M urea; 5 % β-mercaptoethanol; 5 % NP-40; 0.2 % ampholytes
- O-buffer: 0.06 M Tris-Cl, pH 6.8; 2.5 % SDS; 5 % β-mercaptoethanol; 20 % glycerol
- Acidic chamber buffer: 0.01 M H_3PO_4
- Basic chamber buffer: 0.02 M NaOH
- Solution K: 6 M urea; 1 % ampholytes; 5 % NP-40
- Acrylamide solution: 28.4 % acrylamide; 1.6 % bisacrylamide
- Gel solution: 9.2 M urea; 4 % acrylamide solution; 2 % NP-40; 2 % ampholytes
- 20 µl TEMED

– 20 µl 10 % APS
– 1 % agarose in O-buffer

Procedure

Preparation of IEF and NEpHGE Gels

1. Seal glass tubes properly at one end with parafilm and carefully pour the gel solution (avoid air bubbles). All tubes should be poured to the same level (mark in advance). Overlay the gel solution with distilled H_2O and allow to polymerize for several hours or overnight.

2. Rinse gels after polymerization properly with water and put glass tubes into the gel apparatus. Make sure that the apparatus is not leaky; agarose can be used to seal the glass tubes into the gel apparatus. Degas and fill the lower chamber buffer with H_3PO_4 (IEF) or NaOH (NEpHGE) and remove air bubbles under the gel using a syringe.

Prerun of IEF Gel

1. **Do not load sample on the gel for this step!** Overlay gel with solution K (20 µl) and fill upper reservoir with the upper chamber buffer (NaOH).

2. Prerun gel at 200 V for 15 min, than at 300 V for 30 min and at 400 V for 30 min.

3. Remove upper chamber buffer and solution K, load samples on the gel. **Note:** in case of NEpHGE, no prerun is required!

Preparation of Samples

1. Dissolve protein sample in 30 µl lysis buffer I and incubate for 10 min at room temperature.

2. Add 30 µl lysis buffer II and incubate again 10 min at room temperature. Spin samples for 10 min in an Eppendorf centrifuge at full speed to remove unsoluble material.

3. Use cytochrome c as marker protein; prepare as described above.

Electrophoresis

1. Load samples carefully into the glass tubes. Overlay each probe with 20 μl buffer K and afterwards with NaOH (IEF) or H_3PO_4 (NEpHGE).

2. Fill upper chamber with the recommended buffer.
   - Electrodes:
   IEF: Upper: cathode; lower: anode
   NEpHGE: Upper: anode; lower: cathode

3. Running of gel:
   - IEF: Run gel at 400–500 V for 5000–10 000 Vhours.
   - NEpHGE: 10 min at 200 V, 10 min at 300 V, then 400 V until cytochrome c is visible at the lower end of the gel.

4. Press round gels out of the glass tube using a large syringe filled with water. Be careful!! Lay gel on a petri dish, mark one end (H^+ or OH^-). **Note:** If you cannot press the gels out of the glass tube, freeze the gels for a few minutes.

5. Incubate gels in O-buffer for 10 min. Remove O-buffer and store gels at −20 °C, or use them immediately for separation in the second dimension.

6. Cut the cytochrome c gel into 0.5 cm pieces. Add bidistilled water. Shake 1 h at room temperature and monitor the pH of each fraction to analyze the gradient.

7. Second dimension: Put round gel on the stacking gel of the SDS gel and seal it with 1 % agarose in O-buffer with bromophenol blue. For SDS electrophoresis, see section on SDS-PAGE. Make sure that you can differentiate between the H^+ and the OH^- ends of the gel. Note: Add SDS und β -ME after boiling the agarose!

The scheme of a 2-D gel is shown in Fig. 1.5, two-dimensionally separated hnRNP proteins are shown in Fig. 1.6.

Troubleshooting

- Problem: polacrylamide gel does not polymerize.
  - The APS and/or its 10 % solution are not more active. Use only new batches of APS and always prepare fresh 10 % solution. If this does not help, increase the amount of the 10 % solution in the gel solution up to threefold.

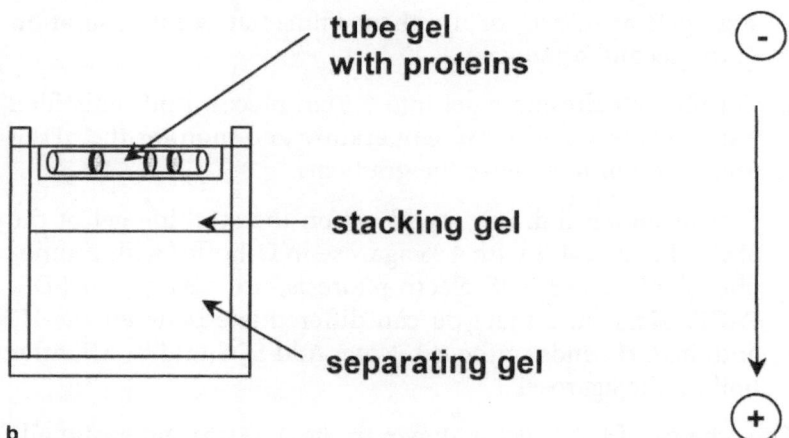

Fig. 1.5a–c. Two-dimensional gel electrophoresis (NEpHGE). **a** First dimension, **b** second dimension

Marker proteins   protein spots

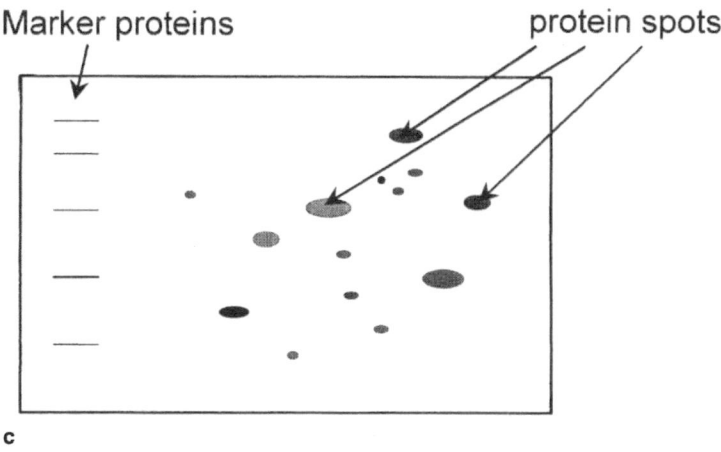

c

Fig. 1.5c. Coomassie-stained gel

- The acrylamide/bisacrylamide solution polymerizes before pouring the gel.
  - The temperature of the solution is too high and is increasing the speed of polymerization. Mix the components of the solution in a vessel cooled in an ice bath.
- After adding the protein sample buffer to the TCA-precipitated protein specimen, the colour of the solution changes from blue to green-yellow.
  - Although the precipitate was washed with acetone, some TCA may have remained in the pellet. Apply gaseous ammonia using a Pasteur-pipette onto the surface of the specimen until its colour changes from green-yellow to blue.
- After staining the protein gel, the proteins appear as a smear instead of distinct bands.
  - The protein samples were not denatured before being applied to the gel. Boil the samples in protein sample buffer at 100 °C for 5 min. Spin for 10 min before loading onto the gel.
- After staining, fingerprints and similar traces appear on the protein gel.
  - Before pouring the gel, clean the glass plates thoroughly with water and dishwashing liquid. Do not touch the gel without gloves.

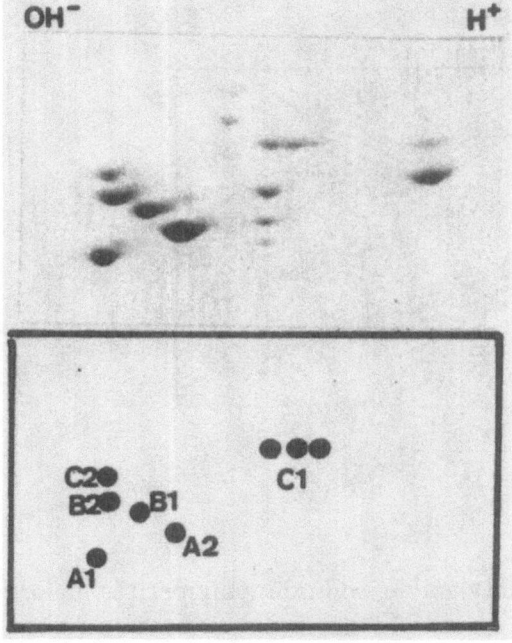

Fig. 1.6. Two-dimensionally separated (NEpHGE/SDS-PAGE) proteins of a 40S peak of HeLa hnRNP

- No pH gradient was formed; no separation by the IEP.
  - Check buffers used. Make sure that the electric field has the right direction. Use new batch of ampholytes.

- Part of the material loaded onto the pH gradient gel does not separate in the second dimension.
  - The pH gradient gel is overloaded and/or the samples are not properly dissolved. Use more sample solution, spin before loading onto the gel.

- Precipitation of urea in the polymerized gel.
  - Room temperature is not high enough.

1.5
Peptide Maps

The two-dimensionally separated proteins can be analyzed by mapping their peptides: The spots can be cut out of the gel and separated after treatment with a sequence-specific protease in a third SDS gel (Cleveland et al. 1977).

Materials

- Peptide mapping buffer: 125 mM TrisHCl, pH 6.8; 1 mM EDTA; 0.1 % SDS

Procedure

1. Stain gel of the second dimension very shortly with Coomassie blue R-250 using freshly prepared solutions. Do not stain longer than 15 min!

2. Destain as shortly as possible as described above, changing the destaining solution frequently. If protein spots are visible, rinse gel 20 min with distilled H_2O. **Note:** Methanol and acetic acid may denature proteins irreversibly!

3. Cut spots out on a transilluminator using a scalpel.

4. Equilibrate gel spots in peptide mapping buffer at room temperature, 15 min.

5. Prepare SDS gel; add EDTA to a final concentration of 1 mM. Prepare stacking gel with 4.5 % acrylamide and 1 mM EDTA.

6. Put equilibrated spots into gel slots, overlay with peptide mapping buffer (with 20 % glycerol).

7. Overlay spots with native protease (dissolved in peptide mapping buffer).

8. Overlay protease with peptide mapping buffer (with 10 % glycerol).

9. Run gel as described, pausing for 30 min if the bromophenol blue front is on the border between the stacking and separating gels. Run gel, make silver stain.

Troubleshooting

- No peptides detected.
  - Check activity of protease and amount of protein to be analyzed.

1.6
Immunoblot (Western Blot)

The proteins separated by SDS gel electrophoresis can be transferred to either nitrocellulose or a nylon membrane. Antibodies directed to these proteins are allowed to react with the separated, denatured proteins. The antibody binding proteins are stained by an appropriate staining assay (Towbin et al. 1979). If the antibody is directed against native epitopes, the proteins separated by SDS-PAGE must be renatured prior to reaction with the antibody. This technique is also a tool to investigate the binding properties of proteins to nucleic acids (Schenkel et al. 1988).

Materials

- Transfer buffer: 25 mM Tris; 200 mM glycine; 20 % methanol
- TBS: 50 mM Tris-HCl, pH 7.4; 200 mM NaCl
- Urea/MgCl$_2$: 3 M urea; 100 mM MgCl$_2$
- Renaturation buffer for gels: 4 M urea; 10 mM Tris-HCl, pH 7.0; 2 mM EDTA; 0.1 mM DTT; 50 mM NaCl
- Blocking solution: 50 % horse serum in TBS
- Ponceau S staining solution: 0.2 % Ponceau S in 3 % TCA
- AP buffer: 100 mM Tris-HCl, pH 9.5; 100 mM NaCl; 5 mM MgCl$_2$
- Stop solution: 20 mM Tris-HCl, pH 8; 5 mM EDTA
- 4-Nitrobluetetracoliumchloride (NBT): 50 mg/ml in 70 % dimethylformamide (DMF)
- 5-Bromo-4-chloro-3-indoxylphosphate (BCIP): 50 mg/ml BCIP in 100 % DMF

Procedure

Protein Transfer

1. Run SDS-PAGE. **Do not stain the gel!**

2. Cut nylon membrane or nitrocellulose and two sheets of Whatman 3MM paper to the size of the gel.

3. Wet membrane carefully with bidistilled water.

4. Make a transfer sandwich without air bubbles, as shown in Fig. 7.5.

5. Put sandwich into the blot chamber filled with refrigerated transfer buffer (**membrane to anode!**).

6. Transfer overnight: 50 V at 4 °C.

7. Renature proteins with urea/MgCl$_2$ after the transfer. **Alternatively,** renature proteins in gel: Incubate gel in renaturing buffer 5 h. Shake gently at room temperature. The gel will become ca. 20 % larger. Measure size of gel. Transfer without methanol.

8. Stain membrane with Ponceau S for 1 min to show the quality of the gel and the transfer. Destain membrane with TBS.

Antibody Reaction

1. Saturate membrane with 50 % horse serum in TBS at room temperature for 1 h.

2. Add first antibody for 75 min in 3 % horse serum in TBS at room temperature.

3. Wash: 3 times for 1 min in 0.1 % Tween in TBS, followed by 3 times for 1 min in TBS.

4. Incubate the second antibody as described above. Equilibrate blot strip after washing in AP buffer for 1 min.

5. Develop blot strip. Add 66 µl NBT solution and 33 µl BCIP solution to 10 ml AP buffer (freshly prepared). If bands are visible, stop reaction immediately.

6. Wash filters with H$_2$O.

7. Store blots in water, take photographs.

For further details see Chap. 7.

Troubleshooting

Transfer of Proteins to Nitrocellulose

- After transfer, no or only weak protein bands are visible on the Ponceau S-stained blotting membrane, whereas protein bands are visible on the Coomassie-stained gel.

- The transfer was performed at too low power or for too short of a time. Increase the power and/or the time of transfer. Check the fuse of the power supply and the quality of the transfer buffer.

- After transfer, protein bands are not visible on the Ponceau S-stained blotting membrane or on the Coomassie-stained gel.
  - The transfer was performed at too high power or for too long, resulting in migration of the proteins through the membrane. Decrease the power and/or the time of transfer.

- After transfer on the Ponceau S-stained membrane, the protein bands are interrupted by circular white areas.
  - While assembling the transfer sandwich, air bubbles were trapped between the gel and the blotting membrane. Squeeze them out using a glass pipette or a glass rod.

- The gel melted during the transfer.
  - The blotting chamber was not cooled during transfer. Perform the transfer at 4 °C and stir the transfer buffer continuously.

Detection of Blotted Proteins

- After developing the blot with NBT and BCIP, there is intensive background staining of the membrane.
  - The nonspecific binding sites of the blotting membrane were not blocked efficiently. Increase the percentage of horse serum in the blocking buffer up to 50 %. Wash the membrane between application of the first and the second antibodies thoroughly.

- While developing the blot with NBT and BCIP, no or only weakly stained bands appear.
  - The second antibody, or one or both of the substrates, does not work anymore. Test the substrate solutions and use a fresh aliquot of the second antibody.

1.7
Immunoprecipitation

Alternatively to Western blots, native proteins are allowed to react with an antibody-protein A-sepharose complex. This complex (antigen-antibody-protein-A-sepharose) is precipitated and the antigen components can be analyzed (immunoprecipitation).

Antigen-antibody complexes can be purified if they are bound to a matrix which binds antibodies. Such a matrix is Sepharose conjugated with protein A, an antibody-binding protein isolated from the membrane of *Staphylococcus aureus*. In immunoprecipitation, the protein is bound to the antibody in a native form. The conditions of the experiment can be changed by variations in the salt concentration.

Materials

- NET-2: 150 mM NaCl; 10 mM Tris-HCl pH 7.5; 0.05 % NP-40
- Stop buffer: 0.1 M glycine pH 3; 0.15 M NaCl

Procedure

Binding of Antibodies to Protein-A-Sepharose

1. Incubate 2.5 mg protein A-Sepharose (PAS) 1 h in NET-2.

2. Spin for 1 min full speed in an Eppendorf centrifuge.

3. Dissolve in 500 µl NET-2. Add antibody and incubate for 1 h at room temperature; shake gently.

4. Spin 1 min full speed in an Eppendorf centrifuge.

5. Wash three times with NET-2 (dissolve in 750 µl NET-2, shake gently for 5 min, spin for 1 min full speed in an Eppendorf centrifuge).

6. Pellet is the PAS-antibody complex (PAS-AB).

Antigen Preadsorbtion with Protein A

1. Obtain the same volume of the commercially available 10 % insoluble protein A suspension as the volume of the antigen sample.

2. Spin for 1 min full speed in an Eppendorf centrifuge.

3. Dissolve in one volume NET-2.

4. Add antigen sample.

5. Incubate 20 min at 4 °C.

6. Spin full speed in an Eppendorf centrifuge to obtain insoluble protein A preadsorbant in the pellet.

Binding of the Antigen to PAS-AB

1. Dissolve PAS-AB in NET-2 and add antigen.

2. Add NET-2 to a final volume of 500 µl.

3. Incubate 1 h at 4 °C.

4. Spin for 1 min full speed in an Eppendorf centrifuge.

5. Wash four times with NET-2. Spin after the last washing step for 4 min and remove supernatant.

6. Pellet is the PAS-antibody-antigen complex (PAS-AB-AG).

Elution of the Immunoprecipitate

1. Dissolve PAS-AB-AG pellet in 100 µl stop buffer.

2. Incubate 5 min at 4 °C.

3. Spin 4 min full speed in an Eppendorf centrifuge.

4. Collect supernatant.

5. Neutralize supernatant with 7 µl 1 M Tris-HCl, pH 8, immediately.

6. Analyze immunoprecipitate by SDS-PAGE.

References

Ares M Jr, Weiser B (1995) Rearrangement of snRNA structure during assembly and function of the spliceosome. Prog Nucleic Acid Res Mol Biol 50:131–159

Biamonti G, Riva S (1994) New insights into the auxiliary domains of eucaryotic RNA binding proteins. FEBS Lett 340:1–8

Burd CG, Dreyfuss G (1994) Conserved structures and diversity of functions of RNA-binding proteins. Science 265:615–621

Cleveland DW, Fischer SG, Kirschner MW, Laemmli UK (1977) Peptide mapping by limited proteolysis in sodium dodecyl sulfate and analysis by gel electrophoresis. J Biol Chem 252:1102–1106

Dreyfuss G, Matunis MJ, Pinol-Roma S, Burd C (1993) hnRNP proteins and the biogenesis of mRNA. Annu Rev Biochem 62:289–321

Görlach M, Burd CG, Portman DS, Dreyfuss G (1993) The hnRNP proteins. Mol Biol Rep 18:73–78

Hodges PE, Beggs JD (1994) RNA splicing. U2 fulfils a commitment. Curr Biol 4:264–267

Hodges D, Bernstein SI (1994) Genetic and biochemical analysis of alternative RNA splicing. Adv Genet 31:207–281

Izaurralde E, Mattaj IW (1995) RNA export. Cell 81:153–159

Laemmli UK (1970) Cleavage of structural proteins during the assembly of the head of bacteriophage T4. Nature (Lond) 227:680–685

Lamm GM, Lamond AI (1993) Non-snRNP protein splicing factors. Biochim Biophys Acta 1173:247–265

Lamond AI (1993) The spliceosome. Bioassays 15:595–603

Louis C, Sekeris CE (1976) Isolation of informoferes from rat liver. Effects of alpha-amanitin and actinomycin D. Exp Cell Res 102:317–328

Lührmann R (1990) Functions of U-snRNPs. Mol Biol Rep 14:183–192

Lührmann R, Kastner B, Bach M (1990) Structure of spliceosomal snRNPs and their role in pre-mRNA splicing. Biochim Biophys Acta 1087:265–292

Mattaj IW, Nagai K (1995) Recruiting proteins to the RNA world. Nat Struct Biol 2:518–522

McEwen CR (1967) Tables for estimating sedimentation through linear concentration gradients of sucrose solution. Anal Biochem 20:114–149

Madhani HD, Guthrie C (1994) Dynamic RNA-RNA interactions in the spliceosome. Ann Rev Genet 28:1–26

Nagai K, Oubridge C, Ito N, Avis J, Evans P (1995) The RNP domain: a sequence-specific RNA-binding domain involved in processing and transport of RNA. Trends Biochem Sci 20:235–240

Newman AJ (1993) RNA:RNA interactions in the spliceosome. Mol Biol Rep 18:85–91

Newman A (1994a) Small nuclear RNAs and pre-mRNA splicing. Curr Opinion Cell Biol 6:360–367

Newman A (1994b) RNA splicing. Activity in the spliceosome. Current Biol 4:462–464

Nigg EA, Baeuerle PA, Lührmann R (1991) Nuclear import-export: in search of signals and mechanisms. Cell 66:15–22

Nilsen TW (1994) RNA-RNA interactions in the spliceosome: unraveling the ties that bind. Cell 78:1–4

O'Farrell PZ, Goodman HM, O'Farrell PH (1977) High resolution two-dimensional electrophoresis of basic as well as acetic proteins. Cell 12:1133–1141

Schenkel J, Sekeris CE, Alonso A, Bautz EK (1988) RNA-binding properties of hnRNP proteins. Eur J Biochem 171:565–569

Swanson MS (1990) Heterogeneous nuclear ribonucleoprotein complexes. Mol Biol Rep 14:79–82

Towbin H, Staehelin T, Gordon J (1979) Electrophoretic transfer of proteins from polyacrylamide gels to nitrocellulose sheets: procedure and some applications. Proc Natl Acad Sci USA 76:4350–4354

Will CL, Behrens SE, Lührmann R (1993) Protein composition of mammalian spliceosomal snRNPs. Mol Biol Rep 18:121–126

Wittop Koning TH, Schümperli D (1994) RNAs and ribonucleoproteins in recognition and catalysis. Eur J Biochem 219:25–42

The lnRNP Particle – A Naturally Assembled Complex of Pre-mRNA and Splicing Factors

RUTH SPERLING[1] AND JOSEPH SPERLING[2*]

Introduction

The processing of nuclear pre-mRNA is an important step in regulating the expression of genes transcribed by RNA polymerase II. These processing events include capping, splicing, 3'-end processing and transport from the nucleus to the cytoplasm. Transcripts of RNA polymerase II are assembled during transcription with proteins and other components into a ribonucleoprotein (RNP) complex and remain associated in RNP particles throughout their residence in the nucleus (Miller and Hamkalo 1972; Sommerville 1981; Wu et al. 1991). It can be logically assumed that the processing events occur within these nuclear RNP complexes. Therefore, the isolation and characterization of RNP particles that package intact nuclear RNA should shed light on the characteristics of the pre-mRNA processing machinery of the living cell.

Interactions of pre-mRNA with specific proteins, and its tertiary folding in RNP complexes, presumably play an important role in nuclear RNA processing. However, information regarding the structure of these complexes and the interactions occurring therein, is rather scarce. We have thus developed a method for the release into the nucleoplasm of nuclear RNP complexes in intact form; these can subsequently be used for biochemical and structural studies (Arad-Dann et al. 1987; Miriami et al. 1994; Spann et al. 1989; Sperling et al. 1985). The method enables the quantitative release (over 85 %) of RNP particles into the nucleoplasm, as confirmed by analysis of specific transcripts (Spann et

* Corresponding author: Joseph Sperling: Tel.: (972)-8-934-2509;
  Fax: (972)-8-934-4142; e-mail: cosper@weizmann.weizmann.ac.il
[1] Department of Genetics, The Hebrew University of Jerusalem,
  Jerusalem 91904, Israel
[2] Department of Organic Chemistry, The Weizmann Institute of Science,
  Rehovot 76100, Israel

al. 1989; Sperling et al. 1985). The released particles sediment as a relatively narrow band at the 200S region in sucrose gradients. In the electron microscope they show a compact structure composed of several substructures with an overall diameter of 50 nm (Miriami et al. 1995; Spann et al. 1989). We thus call them large nuclear RNP (lnRNP) particles (Sperling and Sperling 1990). The unique and most intriguing feature of the lnRNP particles, determined experimentally, is that their size and hydrodynamic properties are independent of the length of, and the number of introns in, the RNA they package (see Table 2.1).

Most of the nuclear polyadenylated RNA is found packaged in lnRNP particles. In addition to pre-mRNA, all five small nuclear RNP complexes (U1, U2, U5 and U4/U6 snRNPs) required for pre-mRNA splicing in mammalian cells (Krämer 1995; Moore et al. 1993; Steitz et al. 1988) are found packaged in 200S lnRNP particles (Miriami et al. 1995; Sperling et al. 1986). Among the non-snRNP spliceosomal proteins which have been distinguished as essential splicing factors, U2AF (U2 snRNP auxiliary factor, which binds to the polypyrimidine tract near the 3' splice site and is required for the binding of U2 snRNP to the branch site; Zamore and Green 1991; Zamore et al. 1992; Zhang et al. 1992), PTB (another polypyrimidine binding protein; Garcia-Blanco et al. 1989; Ghetti et al. 1992; Patton et al. 1991), an 88 kDa essential splicing factor (Ast et al. 1991), and the family of serine/arginine rich (SR) proteins (Fu 1995; Mayeda et al. 1992; Roth et al. 1991; Zahler et al. 1992) have been shown to be integral components of lnRNP particles. Regarding the latter, it is pertinent to note that all the nucleoplasmic phosphorylated SR proteins, which are identified by monoclonal antibody (Mab) 104 (Roth et al. 1990, 1991), are packaged in lnRNP particles (Yitzhaki et al. 1996; for summary and references see Tables 2.1, 2.2). Taken together, our observations support the notion of a unitary structure for lnRNP particles and are consistent with the idea that the lnRNP particle can serve as a suitable mold for the splicing of multi-intron pre-mRNAs in the cell's nucleus. Furthermore, structural studies by automated electron tomography confirmed the uniformity of the lnRNP particles and showed that they are composed of four similar subunits, presumably 60S spliceosomes (Sperling et al. 1997).

Table 2.1. Precursor messenger RNAs packaged in large nuclear RNP (lnRNP) particles

Nuclear RNA	Cells	Size		References
		pre-mRNA	mRNA	
CAD	Syrian hamster	25 000	7 900	Sperling et al. (1985)
DHFR	Syrian hamster	36 000	1 600	Spann et al. (1989)
β-Actin	Syrian hamster; HeLa	5 000	1 800	Sperling and Sperling (1990)
Histone H4	Syrian hamster; HeLa		400	Spann et al. (1989)
Poly(A)$^+$ nuclear RNA	Syrian hamster	Heterogeneous		Spann et al. (1989)

Isolation of Large Nuclear RNP Complexes from Mammalian Nuclei

In developing the protocol for the isolation of nuclear RNP complexes from mammalian cell nuclei, we have employed two major criteria to assess the quality of the preparations. Thus, by quantitative analyses of specific pre-mRNAs, we verified that: (1) more than 85 % of the nuclear population of the specific RNAs were associated with the particles; and (2) that these RNAs were kept undegraded throughout the isolation procedure. Initially, we utilized tissue cultured cells containing amplified genes as a source of an abundant nuclear RNP, with a view to the possible isolation and analysis of homogeneous material. Thus, SV40-transformed Syrian hamster cells, in which the gene for a multifunctional enzyme CAD (carbamoyl-P-synthetase, aspartate transcarbamylase, dihydro-orotase) is amplified about 200-fold (Padgett et al. 1982; Wahl et al. 1979), were used for development of the protocol. This system proved to be appropriate for studying nuclear RNP complexes since transcripts of the CAD gene in the transformed Syrian hamster cells are as abundant in the nucleus as in the cytoplasm, and they constitute about 1 % of the polyadenylated RNA population (Sperling et al. 1985). Later, the procedure described here was also used to prepare lnRNP particles from HeLa cells (Spann et al. 1989), and it can now be easily adapted to other cell lines or tissues. Presently, the development of highly sensitive methods for the detection of specific

RNAs (e.g., PCR) and proteins allows the analysis of lnRNP components of low abundance.

To protect the RNA within the particles from degradation, the RNP complexes are released from purified nuclei by mild sonication in the presence of several potent RNase inhibitors, including ribonucleoside vanadyl complex (Berger and Birkenmeier 1979). A key element in the protocol for the preparation of lnRNP particles involves a step of preparing clean nuclei. Thus, the procedure includes isolation of cells in isotonic buffer, lysis of the cells in a low ionic strength buffer, and separation of clean nuclei from the cytoplasmic fraction by sedimentation through a glycerol cushion. This part of the protocol may be utilized when a preparation of clean nuclei, devoid of cytoplasmic contamination, is required for purposes other than RNP particles preparation. It should be pointed out, however, that the cytoplasmic fraction obtained by this protocol may be contaminated with unbroken cells, which do not pellet through the glycerol cushion used to separate nuclei from cytoplasmic material. Thus, in case a clean cytoplasmic preparation is required, a different protocol is recommended (Sperling et al. 1985).

The next step involves microsonication of the clean nuclei and precipitation of chromatin. To prevent the nonspecific association of RNP complexes with chromatin and other nuclear components, release of the RNP complexes is done in the presence of tRNA (Sperling et al. 1985). The supernatant, which is enriched with lnRNP particles, is fractionated in a sucrose gradient and the distribution of the specific transcripts across the sucrose gradient is analyzed by hybridization.

Materials

- Vanadyl ribonucleoside complexes solution (VR): The original procedure of Berger and Birkenmeier (1979) has been modified omitting the purine nucleosides from the mixture. This preparation avoids solubility problems and yields a potent reagent.
    - Flush about 200 ml of deionized water with argon (or oxygen-free nitrogen) for 30 min.
    - Dissolve 2.44 g (10 mmol) uridine and 2.43 g cytidine (10 mmol) in 80 ml of the gas-flushed water in a 200 ml beaker covered with aluminum foil or in a 3-neck round bottom flask.

- Dissolve 5.06 g (20 mmol) of $VOSO_4 \times 5H_2O$ in 10 ml of argon-flushed water.
- Pass a constant mild stream of argon through the nucleosides solution and add the $VOSO_4$ solution dropwise. Keep the pH between 6 and 7 by adding in parallel drops of 10 M NaOH (about 5 ml are needed). The solution should remain clear during addition of $VOSO_4$. Formation of a white precipitate below pH 6 should be avoided.
- Adjust to pH 7 with 2 M NaOH and top up to 100 ml with argon-flushed water. Store 5 ml aliquots in tightly closed vials at −20 °C.

- Wash buffer (WB): 125 mM KCl; 30 mM Tris-HCl, pH 7.5; 5 mM magnesium acetate; 0.15 mM spermine; 0.05 mM spermidine; 2 mM VR
- Swelling buffer (SB): 10 mM KCl; 30 mM Tris-HCl, pH 7.5; 5 mM magnesium acetate; 0.15 mM spermine; 0.05 mM spermidine; 2 mM VR
- Glycerol swelling buffer (GSB): 25 % glycerol (v/v); 10 mM KCl; 30 mM Tris-HCl, pH 7.5; 5 mM magnesium acetate; 0.15 mM spermine; 0.05 mM spermidine; 2 mM VR
- Sonication buffer (SONB): 10 mM Tris-HCl, pH 8.0; 100 mM NaCl; 2 mM $MgCl_2$; 0.15 mM spermine; 0.05 mM spermidine; 2 mM VR
- 10 × ST2M (no VR): 0.1 M Tris-HCl, pH 8.0; 1 M NaCl; 20 mM $MgCl_2$
- ST2M: 10 mM Tris-HCl, pH 8.0; 100 mM NaCl; 2 mM $MgCl_2$; 2 mM VR
- 15 % Sucrose ST2M (100 ml): 2.5 M sucrose (17.5 ml); 10 × ST2M (no VR) (10 ml); 0.2 M VR (1 ml); diethyl pyrocarbonate (DEPC)-water (71.5 ml)
- 45 % Sucrose ST2M (100 ml): 2.5 M sucrose (52.5 ml); 10 × ST2M (no VR) (10 ml); 0.2 M VR (1 ml); DEPC-water (36.5 ml)

Note: All solutions containing spermine, spermidine and/or VR should be autoclaved without these ingredients and stored at −20 °C. Spermine, spermidine and/or VR should be added to the defrosted solutions just before use.

- Guanidinium thiocyanate (GTC) buffer: 4 M guanidinium thiocyanate; 25 mM sodium citrate, pH 7.0; 10 % Sarkosyl; 0.1 M β-mercaptoethanol; anti-foam (Sigma). Dissolve guanidinium thiocyanate in the citrate buffer. Millipore or centrifuge at 10 000 rpm for 10 min to remove colloidal insoluble material.

Add 10 % Sarkosyl to 0.5 %, and adjust to pH 7.0. **Before use:** add β-mercaptoethanol to 0.1 M and 1–2 drops of anti-foam.
- CsCl /EDTA: 5.7 M CsCl; 0.1 M EDTA
- PIPES buffer: 0.2 M PIPES, pH 6.4; 2 M NaCl; 0.5 M EDTA
- S1 nuclease buffer: 20 mM sodium acetate, pH 4.6; 0.28 M NaCl; 0.5 mM zinc acetate

2.1
Preparation of lnRNP Particles

Since RNA molecules are labile and RNases are very resistant, several precautions are taken to preserve the integrity of the RNA during lnRNP particle preparations. All steps in this protocol should be performed uninterrupted at 4 °C, using sterile glassware and plasticware. The water used for the preparation of all solutions should be treated for several hours with 0.2 % DEPC and autoclaved. As an extra precaution against RNases, vanadyl ribonucleoside complexes should be freshly added at all steps of the preparation, and the RNase inhibitor RNasin should be added when lysis of the nuclei takes place.

Procedures

Preparation of Clean Nuclei

Syrian hamster PALA-resistant cells (line 165–28; Kempe et al. 1976), are grown at 37 °C in 8.5 cm tissue culture plates as described (Swyryd et al. 1974). For optimal results, cells should be subconfluent (10^7 cells/plate for Syrian hamster cells). A typical procedure is given below for 10^8 cells (10 plates); it may be scaled up either by increasing the number of plates or by using larger (15 cm) plates.

1. Remove plates from incubator and put on ice. All further steps should be performed at 4 °C, with no interruptions, using precooled solutions.

2. Aspirate medium and cover the cells in each plate with 2 ml WB.

3. Aspirate and add to each plate 2 ml WB.

4. Scrape cells with a rubber policeman, and collect cells from 10 plates into a 50 ml plastic conical tube. Wash every five plates with additional 2 ml WB and combine with the cells suspension.

5. Centrifuge in a refrigerated (4 °C) centrifuge at 250 g for 4 min.

6. Aspirate supernatant. Resuspend pellet in remaining supernatant, then in 25 ml WB and centrifuge at 250 g for 4 min.

7. Aspirate supernatant. Resuspend pellet in remaining supernatant and then in 2.5 ml SB.

8. Incubate on ice for 10 min to allow the cells to swell, and homogenize with 20 strokes in a 7 ml Dounce homogenizer (pestle B).

9. Overlay the broken cells on a 2.5 ml GSB cushion in a 15 ml conical tube and centrifuge for 4 min at 750 g and 4 °C to pellet the nuclei.

10. Remove upper (aqueous) layer with a Pasteur pipette. (This fraction contains most of the cytoplasmic RNAs and is designated as the cytoplasmic fraction). Remove the lower (organic) layer. (This layer contains some cellular organelles and little RNA). If analysis of the cytoplasmic fraction is not required, remove all supernatant.

11. Resuspend the pellet of nuclei in the remaining supernatant and then in 2 ml GSB. Add 100 µl of 10 % Triton-X100 and mix thoroughly by pumping up and down a Pasteur pipette (about 20 times).

12. Centrifuge for 4 min at 750 g and 4 °C.

13. Aspirate supernatant. Resuspend pellet of nuclei in remaining supernatant and then in 2 ml of SB.

14. Spin for 4 min at 750 g and 4 °C.

15. Aspirate supernatant. The pellet contains clean nuclei as was confirmed by light and electron microscopy.

Preparation of Nuclear Supernatant Enriched in lnRNP Particles

1. Resuspend pelleted nuclei from 10^8 cells in remaining supernatant and then in 1 ml of SONB.

2. Add 200 units of RNasin and divide into two Eppendorf tubes, 0.5 ml in each.

3. Sonicate for 20 s (two 10 s sonications with an interval of 10 s) at 4 °C (on ice) with a Kontes microsonicator setting 8, maximum power.

4. Add to each tube 20 μl of 50 mg/ml yeast tRNA (final concentration 2 mg/ml), and mix thoroughly by pumping up and down a Pasteur pipette (about 20 times).

5. Centrifuge in an Eppendorf microfuge for 3 min at 12 000 rpm and 4 °C. The supernatant is designated the "nuclear supernatant" and is enriched in lnRNP particles.

Preparation of lnRNP Particles

1. Prepare linear sucrose gradients from 5.5 ml each of 15 % sucrose ST2M and 45 % sucrose ST2M in a sterile 13.2 ml polyallomer tube (SW41 Beckman rotor). For the fractionation of sedimentation markers prepare identical gradients, but without VR.

2. Load 0.5 ml of the nuclear supernatant on each sucrose gradient and centrifuge for a total $\omega^2 t = 2500$ (ω is in krpm; t is in hours). For example: 10 900 rpm for 21 h, or 41 000 rpm for 90 min.

3. On a parallel gradient, made without VR, load two to six A_{260} units of tobacco mosaic virus (TMV) particles dissolved in ST2M without VR. TMV particles, used here as a sedimentation marker, have a sedimentation value of 200S.

4. Collect the gradients in 20 fractions, 0.55 ml each, starting from the bottom and using a puncturing device or an equivalent set-up and a peristaltic pump operating at a flow rate of 1.1 ml/min. Pre-flush all surfaces that come into contact with the sucrose solution with 15 % hydrogen peroxide and then with DEPC water.

5. Read the absorption at 260 nm of the fractions in the TMV gradient to determine the position of the 200S peak fraction (usually around fractions 10–11).

6. Store the fractionated gradient at −20 °C.

Refractionation of lnRNP Particles

For certain experiments it may be necessary to further purify the particles from contaminating material that sediments near the top of the gradient. For example, U snRNPs are present in the nuclear supernatant in excess of what is required for the assembly of lnRNP particles. These free snRNPs sediment at 10S–25S, but their peak trails towards higher sedimentation values. We have found that refractionation of the 200S peak region of the first gradient is sufficient to separate such contaminants from the main peak at 200S, though some material is lost in this process.

1. Combine three or four fractions (1.65–2.1 ml) around the 200S peak region of each gradient. Use the TMV marker, run in a parallel gradient, to determine the 200S peak position.

2. Transfer to a collodion bag attached to a concentration apparatus (Sartorius) and dialyze at 4 °C for 2 h against two changes of 500 ml each of precooled ST2M.

3. Continue the dialysis under reduced pressure until the sample volume is 0.25 ml.

4. Combine the concentrated material from two gradients, load on a 15 %–45 % sucrose ST2M gradient, centrifuge and collect fractions as above.

2.2
Analyses of the Distribution of Specific RNAs Within lnRNP Particles

For the biochemical characterization of specific lnRNP particles it is necessary to determine the distribution of specific components (e.g., specific pre-mRNAs, snRNAs, and protein splicing factors) across all fractions of the gradient. RNAs can be determined by a variety of blot hybridization and RT-PCR techniques described in textbooks and commercial manuals. Here we describe the S1 nuclease mapping technique which we use to obtain information about the distribution of splicing intermediates in sucrose gradients. Proteins are analyzed by Western blot.

For the analyses of the distribution of specific RNAs within lnRNP particles, RNA is prepared from each fraction of the gradient and the specific RNA in each fraction is identified by blot hybridization using DNA or anti-sense RNA probes. For quantitive determinations we use nuclease S1 mapping analyses with single-stranded DNA probes. In that way we determined the number of copies of CAD RNA in the 165–28 Syrian hamster cells, its distribution in the nuclear and cytoplasmic fractions and in the gradient fractions after centrifugation (Sperling et al. 1985). This enabled us to follow the balance of material and to show that about 90 % of the nuclear CAD transcripts were recovered in lnRNP particles.

RNA Preparation

Samples for RNA analysis should be free of proteins and especially of nucleases. For the preparation of RNA from gradient fractions, proteolysis followed by phenol extraction and precipitation in the presence of carrier RNA seems to be satisfactory. In experiments involving quantitation of specific RNA transcripts, it is necessary to determine the quantity of such transcrips in whole cells or in cell fractions in order to keep track of the balance of material. In that case, denaturing proteins by the chaotropic reagent guanidinium thiocyanate (GTC) is preferable. We use a modified version adapted from the protocol originally described by Chirgwin et al. (1979) for the preparation of total cellular RNA by extraction with GTC and purification of the RNA by sedimentation through a CsCl cushion (see below). Extraction with GTC in the presence of acidic phenol (Chomczynski and Sacchi 1987; and commercially available kits) is faster and involves less work. We have found, however, that the recovery of undegraded DNA-free RNA is more quantitative in the first method (described here). This protocol can be modified for the preparation of RNA from cell fractions such as nuclei or nuclear supernatants.

Procedure

Proteinase K Digestion

1. To recover RNA from cytoplasmic, nuclear, or gradient fractions, add SDS to 1 % and yeast tRNA to 50 µg/ml.

2. Add proteinase K to 200 µg/ml and incubate for 30 min at 37 °C.

3. Recover the RNA by extraction with an equal volume of phenol:chloroform:isoamyl alcohol (50:48:2).

4. Add 10 M ammonium acetate to a final concentration of 2 M, and precipitate the RNA with 2.5 volumes of ethanol.

Total RNA from Cells Grown in Tissue Culture

1. Aspirate medium.

2. To each 85 mm plate, add 1 ml GTC buffer; tilt the plate to spread the reagent.

3. Shear the DNA by pipetting or passing the viscous solution several times through a 22-gauge syringe needle to reduce viscosity.

4. Load onto 2–2.5 ml CsCl/EDTA in a cellulose nitrate ultracentrifuge tube (0.5"×2").

5. Centrifuge in a Beckman SW50 rotor at 35 000 rpm for 15 h at 15 °C.

6. Remove the upper phase with a Pasteur pipette leaving the RNA pellet and a layer of 0.5–1 cm CsCl/EDTA.

7. Rinse the walls of the tube with GTC (3×0.5 ml) keeping the RNA pellet covered with a CsCl/EDTA layer.

8. Decant the remaining CsCl/EDTA solution, and cut the tube with a scalpel, leaving the bottom 1 cm with the RNA pellet.

9. Rinse the RNA pellet once with 70 % ethanol on ice. This wash facilitates resuspension of the RNA pellet.

10. Resuspend the RNA in cold TE (10 mM Tris-HCl, 1 mM EDTA, pH 8.0); let stand for 30 min on ice for the RNA to dissolve, and transfer to a microfuge tube.

11. Add one-ninth volume of 3 M sodium acetate, pH 5.5, or one-fourth volume of 10 M ammonium acetate, pH 7.5, and precipitate in 2.5 volumes of ethanol.

12. Store the RNA at $-20\,°C$, either as an ethanol precipitate or in a 1:1 ethanol/TE solution (dissolve pellet in TE and add an equal volume of ethanol).

    Note: One subconfluent plate of Syrian hamster cells ($\sim 10^7$ cells) yields 200–250 µg RNA. To prepare RNA from cell fractions (nuclei or cytoplasm), add 5–10 volumes of GTC and proceed as described.

S1 Nuclease Mapping of CAD RNA

Optimal conditions for S1 nuclease mapping analyses should be determined for each pre-mRNA and probe. Here we describe the procedure we use for analyzing CAD RNA in bulk RNA (total, nuclear, or cytoplasmic) or in RNA prepared from sucrose gradient fractions. Probes are made from the appropriate cloned genomic DNA fragments labeled either at the 3'-end or at the 5'-end with [^{32}P]phosphate. Protocols for the preparation of probes and their labeling are given in various textbooks and manuals.

1. Mix 10–100 µg bulk RNA, or RNA prepared from a sucrose gradient fraction, with 50 ng of the appropriate ^{32}P-labeled CAD probe (10^7 cpm/µg), ensuring at least a fivefold excess of the specific DNA sequences. Precipitate with ethanol.

2. Dissolve the pellet in 4 µl PIPES buffer and add 16 µl deionized formamide (for reproducible results use crystallized formamide).

3. Incubate at 76 °C for 10 min and then hybridize at 56 °C for 24–36 h.

4. Mix rapidly with 9.5 volumes ice-cold S1 nuclease buffer.

5. Add denatured, sonicated salmon sperm carrier DNA to a final concentration of 10 µg/ml, and digest with 100–300 units/ml S1 nuclease at 43 °C for 30 min.

6. Stop the reaction by ethanol precipitation.

7. Dissolve the protected DNA fragments in a gel sample buffer containing 80 % formamide, denature at 95 °C for 3 min and

analyze by electrophoresis in denaturing polyacrylamide/
7 M urea gels and autoradiography. As controls, samples
should be analyzed in the absence of added RNA.

2.3
Analyses of the Distribution
of Protein Splicing Factors Across the Gradient

The distribution of spliceosomal U snRNP proteins and of several non-snRNP splicing factors can be analyzed by slot blot or
immunoblotting of the sucrose gradient fractions with the
appropriate antibodies. Antibodies from a panel of monoclonal
antibodies we have raised against InRNP particles (Offen et al.
1987), as well as antibodies against U snRNP proteins, SR proteins (Roth et al. 1990), U2AF (Zamore and Green 1991) and PTB
(Patton et al. 1991), were utilized (see Table 2.2). Protocols suitable for analyses of several protein splicing factors are described
below. However, for probing phosphorylated SR proteins the
procedure of Zahler et al. (1992) is recommended. Some of the
protein components are frequently found both at the 200S region
of the gradient and at the top of the gradient. In these cases
refractionation (see protocol above) is required for determining
the distribution of such proteins within InRNP particles.

Procedure

Protein Dot Blot Analysis

1. Block a nitrocellulose membrane by gentle shaking for 1 h at
   25 °C in a solution of 10 % low-fat milk in PBS containing
   0.1 % Triton X100.

2. Spot aliquots of 5–10 µl from each fraction of RNP particles
   fractionated on sucrose gradients onto the membrane.

3. Incubate with the appropriate antibody diluted in PBS containing 0.1 % Triton X100 for 1 h at 25 °C.

4. Wash five times for 10 min with 0.25 % low-fat milk in PBS/
   0.1 % Triton X100.

Table 2.2. Components of large nuclear RNP (lnRNP) particles

	Cells	Probe	References
U1 snRNA	Syrian hamster; HeLa	DNA; anti-sense RNA	Sperling et al. (1986)
U2 snRNA	Syrian hamster; HeLa	DNA; anti-sense RNA	Sperling et al. (1986)
U4 snRNA	Syrian hamster; HeLa	DNA; anti-sense RNA	Miriami et al. (1995)
U5 snRNA	Syrian hamster; HeLa	DNA; anti-sense RNA	Miriami et al. (1995)
U6 snRNA	Syrian hamster; HeLa	DNA; anti-sense RNA	Sperling et al. (1986)
hnRNP core proteins: A1, A2, B1, B2, C1, C2	Syrian hamster; HeLa	SLE autoantibodies	Sperling and Sperling (1990)
snRNP proteins	Syrian hamster; HeLa	SLE autoantibodies	Sperling et al. (1986)
56 kDa antigen	Syrian hamster; HeLa	Mmyositis autoantibodies	Arad-Dann et al. (1987)
SR proteins	Syrian hamster; HeLa	MAb104	Yitzhaki et al. (1996)
U2AF	Syrian hamster; HeLa	Anti-U2AF[65]	Yitzhaki et al. (1996)
PTB proteins	Syrian hamster; HeLa	Anti-PTB	Yitzhaki et al. (1996)
2'–5' Oligoadenylate synthetase (2'–5') OASE	Syrian hamster; HeLa	Anti-2'–5' OASE	Sperling et al. (1991)
SF783	Syrian hamster; HeLa	Anti-200S MAb 783	Unpublished
88 kDA (SF 53/4)	Syrian hamster; HeLa	Anti-200S MAb 53/4	Ast et al. (1991)
35 kDA protein	Syrian hamster; HeLa	Anti-200S MAb 15/7	Offen et al. (1987)
32 kDA protein	Syrian hamster; HeLa	Anti-200S MAb 84/3	Offen et al. (1987)
45 kDA protein	Syrian hamster; HeLa	Anti-200S MAb 36	Offen et al. (1987)
45 kDA protein	Syrian hamster; HeLa	Anti-200S MAb 85	Offen et al. (1987)
Nuclear[35]S-labeled proteins	Syrian hamster	SLE autoantibodies	Spann et al. (1989)

5. Incubate for 1 h at 25 °C with the appropriate second antibody (or protein A) conjugated to horseradish peroxidase (Amersham) diluted 1:1000 to 1:5000 in the same buffer as above.

6. Wash with 0.25 % low-fat milk in PBS/0.1 % Triton X100 and detect by chemiluminescence using the ECL commercial kit (Amersham) according to the manufacturer's instructions.

Western Blot Analysis

1. Aliquot 200 µl from each fraction of the sucrose gradient and add 800 µl acetone (precooled to −20 °C).

2. Incubate at −20 °C overnight.

3. Collect the precipitate by centrifugation at 10 000 g for 30 min at 4 °C. Remove and discard the supernatant.

4. Wash pellet with fresh cold acetone. Centrifuge again at 10 000 g for 10 min.

5. Discard the supernatant. Air dry at room temperature (about 10 min).

6. Dissolve each sample in gel sample buffer and electrophorese in SDS polyacrylamide gel (8.75–12 % gel).

7. Transfer the gel-separated proteins electrophoretically onto a nitrocellulose membrane using a semi-dry blot apparatus.

8. Stain the nitrocellulose membrane with Ponceau S and mark the position of the molecular weight markers.

9. Block the nitrocellulose membrane by gentle shaking for 1 h at 25 °C in a solution of 10 % low-fat milk in PBS containing 0.05 % Tween-20.

10. Incubate the membrane with the appropriate antibody diluted in PBS containing 0.05 % Tween-20 for 1 h at 25 °C.

11. Wash four times 5 min with 0.25 % low-fat milk in PBS/ 0.05 % Tween-20.

12. Incubate for 1 h at 25 °C with the appropriate second antibody (or protein A) conjugated to horseradish peroxidase (Amersham) diluted 1:1000 to 1:5000 in the same buffer as above.

13. Wash four times 5 min with 0.25 % low-fat milk in PBS/ 0.05 % Tween-20, and two times 5 min with PBS. Detect by chemiluminescence using the ECL commercial kit (Amersham) according to the manufacturer's instructions.

Analyses of Components Associated with lnRNP Particles

The 200S lnRNP particles were defined according to the distribution of several specific transcripts (CAD, DHFR, β-actin and histone H4; see Table 2.1). Hybridization across the gradient with U snRNA-specific probes revealed co-sedimentation of the five spliceosomal U snRNAs with the 200S lnRNP particles. Similarly, several protein splicing factors were shown with the aid of antibodies to comigrate with the lnRNP particles. To confirm that U snRNPs and protein splicing factors are integral components of the 200S lnRNP particles, indirect immunoprecipitation is utilized. For example, an immunoprecipitation protocol using anti-Sm or anti-U1 snRNP antibodies is detailed below.

Procedure

Immunoprecipitations from Sucrose Gradient Fractions

1. Wash 100 mg of protein A Sepharose beads with several volumes of PBS and incubate with PBS containing 1 % RNase-free bovine serum albumin (BSA) for 10 min.

2. Divide into 20 microfuge tubes (5 mg protein A Sepharose per each gradient fraction).

3. Centrifuge at low speed and discard supernatant.

4. Resuspend the beads in each tube in 100 µl of an antibody solution. Incubate with gentle shaking for 2 h at room temperature followed by overnight incubation at 4 °C.

5. Centrifuge gently and discard supernatant (or save for further experiments if titer remains high enough).

6. Wash three times with 1 ml ST2M buffer and discard supernatant.

7. Aliquot 200 µl from each fraction of the sucrose gradient and add to the respective tube containing the antibody-coated beads. Incubate with gentle shaking at 4 °C for 2 h.

8. Aspirate the supernatant and wash the beads twice with 1 ml ST2M.

9. For the analysis of bound RNA, add to each tube 300 µl ST2M buffer, 2 µl 1 mg/ml yeast tRNA, 40 µl 10 % SDS and 80 µl 10 M ammonium acetate. Extract with phenol:chloroform:isoamyl alcohol (50:48:2) and precipitate with ethanol.

10. For the analysis of bound proteins, add to each tube 60 µl 1.5 × protein sample buffer; incubate at 85 °C for 10 min and run on an SDS/polyacrylamide gel.

For analysis of the nonbound material, analyses of RNA and proteins are performed on the supernatant obtained from each gradient fraction. As a control for the specificity of the immunoprecipitation, similar analyses across the sucrose gradient are performed using nonrelevant antibodies.

References

Arad-Dann H, Isenberg DA, Shoenfeld Y, Offen D, Sperling J, Sperling R (1987) Autoantibodies against a specific nuclear RNP protein in sera of patients with autoimmune rheumatic diseases associated with myositis. J Immunol 138:2463–2468

Ast G, Goldblatt D, Offen D, Sperling J, Sperling R (1991) A novel splicing factor is an integral component of 200S large nuclear ribonucleoprotein (lnRNP) particles. EMBO J 10:425–432

Berger SL, Birkenmeier CS (1979) Inhibition of intractable nucleases with ribonucleoside-vanadyl complexes: isolation of messenger ribonucleic acid from resting lymphocytes. Biochemistry 18:5143–5149

Chirgwin JM, Przbyla AE, MacDonald RJ, Rutter WJ (1979) Isolation of biologically active ribonucleic acid from sources enriched in ribonuclease. Biochemistry 18:5294–5299

Chomczynski P, Sacchi N (1987) Single-step method of RNA isolation by acid guanidinium thiocyanate-phenol- chloroform extraction. Anal Biochem 162:156–159

Fu X-D (1995) The superfamily of arginine/serine-rich splicing factors. RNA 1:663–680

Garcia-Blanco M, Jamison SF, Sharp PA (1989) Identification and purification of a 62,000-dalton protein that binds specifically to the polypyrimidine tract of introns. Genes Dev 3:1874–1886

Ghetti A, Piñol-Roma S, Michael WM, Morandi C, Dreyfuss G (1992) hnRNP I, the polypyrimidine tract-binding protein: distinct nuclear localization and association with hnRNAs. Nucleic Acids Res 20:3671–3678

Kempe TD, Swyryd EA, Bruist M, Stark GR (1976) Stable mutants of mammalian cells that overproduce the first three enzymes of pyrimidine nucleotide biosynthesis. Cell 9:541–550

Krämer A (1995) The biochemistry of pre-mRNA splicing. In: Lamond AI (ed) Pre-mRNA processing. RG Landes Company, pp 35–64

Mayeda A, Zahler AM, Krainer AR, Roth MB (1992) Two members of a conserved family of nuclear phosphoproteins are involved in pre-mRNA splicing. Proc Natl Acad Sci USA 89:1301–1304

Miller OLJ, Hamkalo BA (1972) Visualization of RNA synthesis on chromosomes. Annu Rev Cytol 33:1–25

Miriami E, Angenitzki M, Sperling R, Sperling J (1995) Magnesium cations are required for the association of U snRNPs and SR proteins with pre-mRNA in 200S large nuclear ribonucleoprotein (lnRNP) particles. J Mol Biol 246:254–263

Miriami E, Sperling J, Sperling R (1994) Heat shock affects 5' splice site selection, cleavage and ligation of CAD pre-mRNA in hamster cells, but not its packaging in lnRNP particles. Nucleic Acids Res 22:3084–3091

Moore MJ, Query CC, Sharp PA (1993) Splicing of precursors to mRNA by the spliceosome. In: Gesteland RF, Atkins JF (eds) The RNA World. Cold Spring Harbor Laboratory Press, Cold Spring Harbor, New York, pp 303–358

Offen D, Spann P, Sperling R, Sperling J (1987) Monoclonal antibodies against large nuclear RNP particles. Mol Biol Rep 12:183–184

Padgett RA, Wahl GM, Stark GR (1982) Structure of the gene for CAD, the multifunctional protein that initiates UMP synthesis in Syrian hamster cells. Mol Cell Biol 2:293–301

Patton JG, Mayer SA, Tempst P, Nadal-Gimard B (1991) Characterization and molecular cloning of polypyrimidine tract-binding protein: a component of a complex necessary for pre-mRNA splicing. Genes Dev 5:1237–1251

Roth MB, Murphy C, Gall JG (1990) A monoclonal antibody that recognizes a phosphorylated epitope stains lampbrush chromosome loops and small granules in the amphibian germinal vesicle. J Cell Biol 111:2217–2223

Roth MB, Zahler AM, Stolk JA (1991) A conserved family of nuclear phosphoproteins localized to sites of polymerase II transcription. J Cell Biol 115:587–596

Sommerville J (1981) Immunolocalization and structural organization of nascent RNP. In: Busch H (ed) The Cell Nucleus, vol 8. Academic Press, New York, pp 1–55

Spann P, Feinerman M, Sperling J, Sperling R (1989) Isolation and visualization of large compact RNP particles of specific nuclear RNAs. Proc Natl Acad Sci USA 86:466–470

Sperling R, Sperling J (1990) Large nuclear ribonucleoprotein particles of specific RNA polymerase II transcripts. In: Strauss PR, Wilson SH (eds) The Eukaryotic Nucleus, Molecular Biochemistry and Macromolecular Assemblies, vol 2. Telford, Caldwell, New Jersey, pp 453–476

Sperling R, Sperling J, Levine AD, Spann P, Stark GR, Kornberg RD (1985) Abundant nuclear ribonucleoprotein form of CAD RNA. Mol Cell Biol 5:569–575

Sperling R, Spann P, Offen D, Sperling J (1986) U1, U2 and U6 small nuclear ribonucleoproteins (snRNPs) are associated with large nuclear RNP particles containing transcripts of an abundant gene in vivo. Proc Natl Acad Sci USA 83:6721–6725

Sperling J, Chebath J, Arad-Dann H, Offen D, Spann P, Lehrer R, Goldblatt D, Jolles B, Sperling R (1991) Possible involvement of (2'-5')oligoadenylate synthetase activity in pre-mRNA splicing. Proc Natl Acad Sci USA 88:10377–10381

Sperling R, Koster AJ, Melamed-Bessudo C, Rubinstein A, Angenitzki M, Berkovitch-Yellin Z, Sperling J (1997) Three-dimensional image reconstruction of large nuclear RNP (lnRNP) particles by automated electron tomography. J Mol Biol 267:570–583

Steitz JA, Black DL, Gerke V, Parker KA, Krämer A, Frendewey D, Keller W (1988) Functions of the abundant U-snRNPs. In: Birnstiel ML (ed) Structure and Funciton of Major and Minor Small Nuclear Ribonucleoprotein Particles. Springer, Berlin Heidelberg New York, pp 115–154

Swyryd EA, Seaver SS, Stark GR (1974) N-(Phosphonacetyl)-L-aspartate, a potent transition state analog inhibitor of aspartate transcarbamylase, blocks proliferation of mammalian cells in culture. J Biol Chem 249:6945–6950

Wahl GM, Padgett RA, Stark GR (1979) Gene amplification causes overproduction of the first three enzymes of UMP synthesis in N-(phosphonacetyl)-L-aspartate-resistant hamster cells. J Biol Chem 254:8679–8689

Wu Z, Murphy C, Callan HG, Gall JG (1991) Small nuclear ribonucleoproteins and heterogeneous nuclear ribonucleoproteins in Amphibian germinal vesicle: loops, spheres and snurposomes. J Cell Biol 113:465–483

Yitzhaki S, Miriami E, Sperling R, Sperling J (1996) Phosphorylated Ser/Arg-rich proteins: limiting factors in the assembly of 200S large nuclear ribonucleoprotein particles. Proc Natl Acad Sci USA 93:8830–8835

Zahler AM, Lane WS, Stolk JA, Roth MB (1992) SR proteins: a conserved family of pre-mRNA splicing factors. Genes Dev 6:837–847

Zamore PD, Green MR (1991) Biochemical characterization of U2 snRNP auxiliary factor: an essential pre-mRNA splicing factor with a novel intranuclear distribution. Embo J 10:207–214

Zamore PD, Patton JG, Green MR (1992) Cloning and domain structure of the mammalian splicing factor U2AF. Nature 355:609–614

Zhang M, Zamore PD, Carmo-Fonseca M, Lamond AI, Green MR (1992) Cloning and intracellular localization of the U2 small nuclear ribonucleoprotein auxiliary factor small subunit. Proc Natl Acad Sci USA 89:8769–8773

Intrinsic Fluorescence Techniques for Studies on Protein-Protein and Protein-RNA Interactions in RNP Particles

KARL OTTO GREULICH*

Introduction

The vast majority of newly isolated proteins is often available only in micrograms. Before sacrificing the major part of a valuable preparation for material-consuming experiments, one would like to get as much information as possible from highly sensitive techniques. Analysis by light is particularly material conserving but this technique is quite often not used to its full potential. The purpose of this chapter is to show the potential of simple fluorescence experiments in obtaining preliminary information on a new protein. Once this information is available, more detailed investigations into a number of functional aspects by comparatively simple experiments can then be carried out. In all the experiments described here, intrinsic (natural) fluorescence is used, i.e., the proteins and protein-RNA complexes can be used afterwards in their native form. No staining is necessary. Only experiments which can be performed with a simple steady state fluorescence spectrometer will be described.

The fluorescence of ribonucleoprotein (RNP) particles is governed by the aromatic amino acid tryptophan. Since tyrosine is also a valuable reporter group for interactions, a few experiments with the filamentous bacteriophage Pf1 and with the protamine thynnine, which both reveal tyrosine like fluorescence, will also be reported for comparison.

* Institute of Molecular Biotechnology, Beutenbergerstr. 11,
 Postfach 100813, 07708 Jena, Germany; Tel.: (+49)-3641–65–6400;
 Fax: (+49)-3641–656410; e-mail: KOG@IMB-JENA.DE

Light Scattering, Absorption and Fluorescence Emission

In order to be useful for analysis, light has to be scattered or absorbed. Light scattering is a very short-term process (in picoseconds, 10^{-12} s) which can be used, for example, to get information on the shape of a molecule or a molecular complex. Light scattering will only play a minor role in the analysis of RNP particles.

In contrast, absorption brings a molecule from its ground state into an excited state which, in a mechanical correlate, can be imagined as a spring under tension. After a certain time, often after a few nanoseconds (10^{-9} s), the spring spontaneously relaxes, the molecule de-excites and returns to its ground state. In many cases the energy contained in the excited molecule is converted only into heat and this process is difficult to measure. A certain class of molecules, the fluorescent dyes, however, emit this energy as light, which can be easily measured. Due to the inherent physical mechanisms, the emitted fluorescence light always has a longer wavelength than the exciting light; that is, the emitted light is shifted to red wavelengths. Such "Stokes" shifts may range from a few to more than 100 nm. The simplest fluorescence experiments investigate the wavelength dependence, i.e., they exploit spectral properties. The protocols described below refer to such "steady state" fluorescence studies. Additional information can be (and has been for RNP particles) obtained from time-resolved fluorescence measurements. These, however, will only briefly be mentioned since such measurements require more expensive equipment.

The Aromatic Amino Acids Tryptophan and Tyrosine: Intrinsic UV Fluorescence Probes in Proteins

A typical fluorescence dye absorbs and emits at visible wavelengths, i.e., at wavelengths between 400 nm (violet) and 720 nm (deep red). Such dyes are often used in fluorescence microscopy. A few common dyes, however, absorb in the ultraviolet and emit in the visible range of the spectrum, for example, a number of dyes staining DNA or RNA. A commonly used dye is the type contained in laundry powders. It attaches to textiles and absorbs invisible UV light which is Stokes-shifted and re-emitted as visible light. The human eye, insensitive to UV, perceives this as the generation of light from the dark. Therefore, occasionally textiles appear to be "whiter than white".

A third group of fluorescence dyes absorb and emit in the ultra-violet, i.e., at wavelengths below 400 nm. For the naked human eye nothing appears to happen. However, with suitable detectors, this process can also be observed. It is exactly this group of dyes which is of interest in the context of RNP research. The aromatic amino acids trytophan and tyrosine, and to a minor extent also phenyl-alanine, are representatives of this group.

While the experimental equipment for UV dyes is somewhat more sophisticated than for visible dyes, the advantages of intrinsic (natural) dyes such as tryptophan and tyrosine is obvi-ous: As already mentioned, no staining with extrinsic (foreign) fluorescence dyes is required, which might modify the molecular properties and disturb molecular interactions or which might have low staining yields, thus requiring large amounts of material.

Fluorescence Properties of Tryptophans and Tyrosines in Proteins

On average, over all known proteins every 26th amino acid is a tyrosine and every 68th is a tryptophan. Therefore, large pro-teins will probably contain both amino acids, but smaller pro-teins may have only tyrosine. In order to understand how the aromatic amino acids determine the fluorescence properties of proteins and protein-nucleic acids complexes, one has to know the optical properties of the optically relevant molecules. Ta-ble 3.1 lists some of these properties for tyrosine and trypto-phan. For comparison, phenylalanine and nucleic acids are also included in the list.

Table 3.1. Optical properties of some amino acids and nucleic acids at room temperature in pH 7

	Absorption coefficient[a]	Absorption wavelength (nm)	Fluorescence maximum (nm)	Fluorescence yield[b]
Trp (in water)	5 600	280	355	0.16
Trp (non-polar)	5 600	280	330	0.21
Tyr	1 400	274	305	0.065
Phe	200	257	282	0.04
Nucleic acids[c]	>10 000	260	320	0.0002

[a] See next page.
[b] The fluorescence yield multiplied by 100 is the percentage of absorbed intensity which is finally re-emitted as fluorescence intensity.
[c] Approximate values; definite value depends on base composition.

Nucleic acids at a wavelength of 260 nm are good absorbers, but their fluorescence yield at room temperature is very poor. Therefore they are only indirectly (see below) suitable for fluorescence studies. Phenylalanine is also not very suitable, because of its poor absorption. Tyrosine is a reasonable absorber and has a reasonable fluorescence yield. Tryptophan is the best fluorescence probe due to its quite high absorption coefficient and its good fluorescence yield. In addition, tyrosines tend to transfer their fluorescence energy to tryptophans (see for example Stryer 1978). Therefore, most proteins containing at least one tryptophan reveal the fluorescence spectrum dominated by tryptophan.

Some Quantitative Relationships for the Molar Absorption Coefficient

When light with the intensity I_0 falls on an x cm thick cuvette which is filled with a protein or nucleic acid solution with the concentration c (in moles/liter), the intensity I after the cuvette is given by:

$$I(x) = I_0 \times e^{-\varepsilon \times c \times x} \tag{1}$$

where ε is the molar absorption coefficient in liters/mole and cm (it is a material constant; see Table 3.1). This is the Lambert-Beer law.

In most standard cuvettes x is 1 cm. The exponent $\varepsilon \times c \times x$ is often called the absorption with the symbol A. Note that this equation is occasionally given in a different form:

$$I = I_0 \times 10^{-\alpha \times c \times x} \tag{2}$$

Both equations describe the same fact. The correlation between α and ε is:

$$\alpha = \varepsilon/2.30. \tag{3}$$

The exponent $\alpha \times c \times x$ is occasionally called the OD value.

For example, at 274 nm a solution of 1 mmol/l tyrosine in a 1 cm cuvette causes an absorption of 1.4 (ε for tyrosine is 1400 liters per mole and centimeter, $c=0.001$ mol/l and $x=1$ cm, i.e., $\varepsilon \times c \times x = 1.4$). According to Eq. (1), an absorption of 1.4 means that an intensity I_0 entering the cuvette will be reduced to $I_0/e^{1.4}$.

This is approximately $I_0/4$, as can be verified with any pocket calculator equipped with the exponential function.

Correspondingly, 1 μmol/l would cause an absorption of 0.0014. An absorption of 0.0014 means that the intensity after the cuvette is $I_0/e^{0.0014}$ or 99.8 % of the incoming intensity. Note that only 1 ml of material is needed for an experiment, i.e., the total amount required is of the order of a nanomole. Since 1 mol tyrosine is 181 g, a fraction of a microgram is sufficient for an experiment.

What Questions Can Be Answered by Intrinsic Fluorescence Studies

A number of different questions can be tackled by fluorescence studies, The most simple application of the fluorescence technique is to check the quality of a preparation. For example, when a molecule is not expected to contain tryptophan, an impurity of the latter is easily detected via the fluorescence spectrum. In turn, when the molecule is expected to contain tryptophan, a spectral shift (see Table 3.1) of the fluorescence maximum might indicate denaturation or degradation. If the structure of the molecule is not yet known, there are several fluorescence checks which can indicate whether tyrosine or tryptophan is in the interior of the molecule or exposed to solvent. By adding molecules to the solvent which can reduce (quench) fluorescence, the microenvironment of exposed aromatic amino acids can be elucidated. When the fluorescence of such an exposed group is quenched only by positively charged quencher molecules, this indicates a negative microenvironment.

In a similar way information can be obtained on protein-protein or on protein-nucleic-acid interactions. This is particularly useful in studies of RNP particles. Finally, using fluorescence energy transfer techniques, pairs of fluorescing molecules can be used as intramolecular "rulers" to measure distances of the order of a few angstroms.

▨ Outline

All steady state fluorescence measurements can be performed according to the following protocol:

1. Dissolve the molecule or molecular complex in 1–1.5 ml solvent.

2. Switch the excitation wavelength of the steady state fluorescence spectrometer to 280 nm.

3. Search for the optimal fluorescence emission wavelength.

4. Add modifying, nonfluorescent substances (such as salts or denaturing agents).

5. Measure fluorescence. Evaluate.

▨ Materials

Equipment

For most practical experiments in RNP research a simple fluorescence spectrometer is sufficient. Such a spectrometer is in principle a light source with variable wavelength which allows excitation of the fluorescence. A detector observes the light emitted from the sample at an angle of 90° with respect to the excitation light. Since the major intrinsic fluorescing amino acids (fluorophores) are tryptophan and tyrosine, which are excitable in the ultraviolet (in UV B), the spectrometers have to be equipped correspondingly.

A photometer or an absorption spectrometer

Very helpful, though not absolutely essential, is equipment for measuring the absorption of the sample, i.e., an absorption spectrometer or a photometer. The OD value should be below 0.05 in order to avoid reabsorption of emitted fluorescence light. A simple form of this kind of equipment is a photometer which measures the absorption only at a preselected wavlength, i.e., 260 nm (nucleic acids absorption in protein-nucleic acids complexes) and 280 nm (tyrosine and tryptophan absorption). When only fluorescence studies with RNPs are planned, the cheaper photometer is even better than an absorption spectrometer, which allows measurements at variable wavelength, since with the latter slight irreproducibilies may arise from errors in wavelength setting. In turn, the absorption spectrum and its mathematical derivatives can provide additional information on structures and interactions.

When no photometer or absorption spectrometer is available, linearity can be assessed by measuring the fluorescence as a function of concentration in a dilution series, thereby determining the limits of linearity between concentration and fluorescence intensity.

Cuvettes One should at least have two quartz cuvettes; however, a set of four is recommended. The standard cross section is 1×1 cm with a filling height of 1.5–2 cm, i.e., 1.5–2 ml is required. For smaller volumes, microcuvettes are available with a shorter optical path; but this results in a smaller signal.

Fortunately, fluorescence is a highly sensitive technique, requiring comparably small amounts of material. This, in turn, demands very clean equipment. When quantitative measurements are planned, it is absolutely essential to keep the cuvettes clean. A beginner in fluorescence spectroscopy will be surprised how many materials in the laboratory reveal fluorescence and how persistently they can stick to the surface of cuvettes. Therefore it is highly recommendable to have one reference cuvette which, from the moment of purchase, never comes into contact with nucleic acids, proteins or even with certain solvents (see below). This cuvette should be kept in a very clean container. The sample cuvette(s) have to be checked before each experiment with respect to absorption and fluorescence. Often it is necessary to rewash the cuvette several times until its contaminating signal is smaller than the expected sample signal. The most time is lost by waiting until the cuvettes are perfectly dry. Therefore a cuvette centrifuge, which does not cost more than a few quartz cuvettes, may highly speed up the washing procedure.

While all these precautions at a first glance appear to be tedious, they very quickly become routine and will no longer be perceived as time consuming. The reward for following these strict procedures is that one can work with very small amounts of material (micrograms are ample, even nanogram amounts may be sufficient for a series of experiments) and the time saved for the preparation of the material is much more than the time required for the precaution measures.

Solvents The major solvent is water. Deionized or distilled water should be used but it is usually not necessary to use double distilled water. While it is useful to check occasionally if the water itself contains some fluorescent contamination, this is not too critical. In contrast, a critical aspect is micro-air bubbles which form during the experiment, for example, when the temperature of the sample increases. Thus it is very useful to degas all solvents either by heating up and stirring the water or by using a simple water jet vacuum pump. Surprisingly, some solvents sold as spectroscopically pure may cause problems. For example, some brands of dioxan, a solvent used to mimic a hydrophobic environment, are occasionally stabilized with phenol and the latter has absorption and fluorescence properties which may be confused with those of tyrosine. Thus, when purchasing "spectroscopically pure" solvents one should ask the supplier for which wavelength region the purity is guaranteed.

More than in any other type of experiment, sample preparation can be the main source of errors. In molecular biology a sample often is regarded as pure when contamination (with respect to nucleic acids or contaminating proteins) is <1 %. Constituents of solvents used during preparation may seriously affect the final fluorescence result. For example, when CsCl from a centrifugation gradient is not totally removed, small amounts may cause quenching (see Sect. 3.2) and the fluorescence of amino acid side chains at the surface of a protein may be underestimated. Phenol, already mentioned in the context of solvent quality, may also come from extraction steps during preparation of the sample. Table 3.2 gives an overview of harmful effects of some materials in minor amounts. In large amounts there are many more substances that may ruin the experiments.

The sample

Thus, a sample to be analyzed by intrinsic fluorescence spectroscopy not only has to be pure with respect to large molecules, but also with respect to small solute molecules. If it is not possi-

Table 3.2. Effects of small amounts of some selected substances from the preparation and purification steps

Substance	Effect	Comment
KI	May quench surface groups	Very critical
CsCl	May quench surface groups	Very critical
Acrylamide	May quench surface groups	Critical
SDS	May simulate hydrophobic environment	Less critical
Sugars		Less critical
NaCl		Less critical
Phenol	May simulate tyrosine	Very critical

ble to remove such molecules at all, it is at least necessary to include their possible effects in the discussion of the results.

3.1
Aromatic Amino Acids: Is Tryptophan There?
In Which Microenvironment Is It?

The most simple application of steady state fluorescence is checking the quality of a preparation. For example, when the latter is expected to contain tyrosine but not tryptophan, the fluorescence emission maximum should occur at 305 nm (Table 3.1). When a protein which is known to have tryptophans in an hydrophilic environment, and which should thus fluoresce at 355 nm, shows fluorescence at 330 nm, or vice versa, this may indicate denaturation. The fluorescence emission maximum for tryptophan depends on its environment. In water (corresponding to tryptophan at the surface of a protein) the fluorescence maxima are found at an emission wavelength of 355 nm. In nonpolar solvents (corresponding to the interior of a protein) the emission maximum is around 330 nm. Therefore, just by looking at the fluorescence emission spectrum of a protein one can determine if the tryptophans of a protein are primarily exposed to solvent or are in an hydrophobic microenvironment. One has, however, to be aware that information can be obtained only for those amino acids which govern the fluorescence. For example, in most proteins containing tryptophan nothing can be learned about the tyrosines.

▓ Procedure

1. Rinse quartz cuvette repeatedly and dry carefully, ideally in a cuvette centrifuge. Check against a standard cuvette whether absorption at 280 nm is below 0.005 and whether the fluorescence emission between 305 and 355 nm is much smaller than the expected fluorescence of the sample.

2. Transfer a a few microliters containing RNP particles into a quartz cuvette.

3. Dilute with degassed buffer to an absorption at 280 nm of not more than 0.05.

4. If absorption cannot be reached, repeat experiment by scaling it up or down by a factor of three.

5. Set excitation wavelength to 280 nm. Scan the emission wavelength and look for a maximum.

6. Fix emission wavelength maximum. Scan for excitation wavelength until emission reaches its highest value.

7. Fix emission wavelength. Scan for exciation wavelength until emission reaches its highest value.

8. Optional: For very high sensitivity and accuracy repeat steps 6 and 7.

3.2
Solute Quenching: Tyrosine or Tryptophan – in the Interior or at the surface of a Molecule?

In the previous section, it was shown that, by looking at tryptophan's maximum fluorescence spectrum, one could obtain information on its microenvironment and thus get some idea on its exposure to solvent. However, the fluorescence spectrum is only half of the truth. First, for tyrosine no information on solvent exposure can be obtained from the spectrum, since tyrosine is not sensitive to environment (Table 3.1). Second, even for tryptophan it is difficult to interpret what it means to be in a hydrophobic microenvironment and thus to have a fluorescence maximum at 330 nm (see, for example, Zhang and Hermans 1996). When water molecules are repelled from tryptophan by hydrophobic neighbouring groups, this may be sufficient to cause a hydrophobic microenvironment, in spite of the fact that tryptophan still has some accessibility to solvent. More reliable information regarding accessibility can be obtained by adding solutes which can reduce (quench) the fluorescence (Eftink and Ghiron 1981; Schwarzwald and Greulich 1988). Such solutes may be large ions, for example Cs^+ or I^-, or neutral molecules such as acrylamide. Ions are particularly interesting since they provide additional information on the electric charge of tryptophan's or tyrosine's environment. Only the ions Cs^+ (CsCl) and I^- (KI) will be used in the experiments described below.

In order to understand the principle, imagine a model molecule with only two tyrosines as the single aromatic amino acids:

When a quencher is added and no effect on fluorescence is seen, both tyrosines obviously are buried. The following limiting results are possible:

- Both tyrosines buried: no effect of quencher
- One tyrosine exposed: reduced fluorescence
- Both tyrosines exposed: no residual fluorescence

Even more information can be obtained when the charged quenchers are used. For example, when the negatively charged iodide ion quenches much more efficiently than the positively charged cesium ion one can conclude that the quenched amino acid is in a positively charged microenvironment which attracts the negative quencher and repels the positively charged one.

Such quenching experiments cannot only be performed as qualitative studies – it is also possible to get quantitative answers, i.e., to ask how much a certain amino acid side chain is exposed. For that purpose the Stern-Volmer law can be used:

$$F_o \, / \, F = 1 + K_{SV} \times [Q] \qquad (4)$$

where [Q] ist the quencher concentration in moles/liter, F_o is the fluorescence without quencher and F the fluorescence at various quencher concentrations. K_{SV} (the Stern-Volmer constant) is the slope of the straight line.

This law predicts that a plot of the fluorescence intensity F_o divided by the fluorescence intensity F as a function of quencher concentration will give a straight line. The slope of the Stern-Volmer plot (K_{SV}) for a protein or a RNP particle can be compared with the slope of the free amino acid tryptophan (or tyrosine when the protein reveals tyrosine fluorescence). The ratio of both (a dimensionless number) is a measure of the accessibility.

In the "Results" section it will be shown how K_{SV} can be used to determine the accessibility of those amino acid groups in a protein which govern its fluorescence behaviour.

For very high accuracy a modification of Eq. (4) has to be used, which exploits the lifetime of the fluorescence instead of its intensities. This requires, however, more sophisticated equipment allowing time-resolved fluorescence measurements. In many practical cases (i.e., when all molecules compared in an experiment have similar fluorescence lifetimes), Eq. (4) will give reasonable information on the accessibility of tryptophans or tyrosines.

▒ Procedure

1. Prepare stock solutions of quencher (1 M KI and 5 M CsCl)

2. Fill cuvette with 800 µl of your molecule solution. Measure fluorescence at conditions determined as described above.

3. Add 50 µl quencher stock solution. Cover the cuvette with a Teflon lid. Turn gently several times upside down to mix. Avoid producing air bubbles, even nonvisible ones.

4. Measure fluorescence intensity at identical conditions as above. Calculate F_o/F.

5. Repeat steps 3 and 4 until cuvette is full.

6. Repeat same experiment with CsCl as quencher.

7. Repeat same experiment with the corresponding free amino acid, i.e., with tyrosine or with tryptophan.

8. Calculate relative slopes, i.e., slope of protein or RNP particle divided by slope of free amino acid for both quencher types.

3.3
Is an Aromatic Amino Acid Involved in Protein Nucleic Acid Binding?

Fluorescence cannot only be reduced by a solute quencher but also by a reaction partner. The mechanism for this fluorescence reduction is somewhat different to that described above. The amino acid, tyrosine or tryptophan, emits radiation while the the partner molecule, here a nucleic acid, acts as an antenna which receives the radiation. The efficiency of this energy transfer process depends on the distance between amino acid and nucleic acid, acccording to:

$$dE = r^{-6}/(r^{-6} + R_o^{-6}) \tag{5}$$

where dE is the fraction of energy transferred between the two partner molecules and r is the distances between the ring centers of the aromatic amino acids and the nucleobases of the nucleic acids (for the theoretical basis see Stryer 1979; for a recent sophisticated application see Szmacinski 1996). R_o is the Förster distance for a given pair of emitter and absorber; it can be calculated from molecular properties. For our purposes, it is sufficient to know the Förster distance for the following molecule pairs:

– Tyrosine-nucleic acids: 12.2 Å
– Tryptophan-nucleic acids: 3.4 Å

(More precisely, the Förster distance is somewhat different for the single bases of nucleic acids. The value for total nucleic acids is an average over the four nucleic acids bases.)

Procedure

1. Find optimal fluorescence conditions for your protein-nucleic acid complex as described above.

2. Measure the absolute value of the complexes' fluorescence.

3. Dissociate the protein-nucleic acids complex by adding NaCl (never use salts with large cations or anions since they may quench fluorescence on their own).

4. Measure fluorescence of the dissociated complex.

5. Check on an agarose gel if the complex is really dissociated (lane 1: complex without salt; lane 2: with salt. The latter should give two smaller bands and no complex band).

6. Use Eq. (5) to calculate the distance between the aromatic amino acid and the nucleic acids.

3.4
The Contribution of Electrostatic and Nonelectrostatic Forces to Protein-Nucleic Acid Binding

A major type of interaction of proteins with nucleic acids is the electrostatic (Coulomb) interaction, mediated by positively charged amino acids in the proteins with negatively charged nucleic acids. Generally, a protein-protein, protein-DNA or protein-RNA complex which is stabilized by a combination of electrostatic and other forces will dissociate in high NaCl concentration according to:

$$\log K_{obs} = \log K_o - 0.88 \times m \times \log[NaCl] \tag{6}$$

where [NaCl] is the salt concentration in mol/l; m is the number of electrostatic bonds involved in the process; K_{obs} is the binding constant at a given NaCl concentration and K_o is the binding constant observed at 1 M NaCl. K_o is a measure for nonelectrostatic contributions. A plot of K_{obs} vs the salt concentration on double logarithmic paper (or correspondingly a plot of their logarithms on linear paper) will give a straight line. Its slope divided by 0.88 gives the number of electrostatic bonds involved in the interaction. Its intersection with the vertical axis gives the contribution of the nonelectrostatic interactions.

Fluorescence often changes during salt-induced dissociation of protein-nucleic acid complexes, due to changes in fluorescence energy transfer from proteins to nucleic acids. When salt is added, the fluorescence intensity I increases and finally saturates at a value I_∞. It can be shown that the ratio I/I_∞ is related to K_{obs}. A detailed calculation (Ausio 1984) shows that Eq. (6) then becomes

$$\log\left[(1\text{-}a)/a^2\right] = \log K_o\,[\text{protein}_{total}] - 0.88 \times m \times \log[\text{NaCl}] \quad (7)$$

where $a = I/I_\infty$; and [protein $_{total}$] is the concentration of all protein material used in the experiment.

Apart from its somewhat difficult mathematical derivation, this formula is much simpler to use than it might appear and results in the following protocol:

Procedure

1. Prepare a 3.3 M NaCl solution in deionized water.

2. Determine total protein concentration in the protein-nucleic acids complex solution.

3. Dilute protein-nucleic acid complexes in low salt buffer.

4. Determine optimal fluorescence conditions as described above.

5. Fill 1 ml of the complex in the 2 ml quartz cuvette. (Check if your spectrometer works correctly when only 1 ml is in the cuvette. Otherwise, scale protocol up by a factor of 1.2 or 1.5).

6. Measure fluorescence intensity.

7. Add 100 µl salt stock solution. Cover the cuvette with a Teflon lid. Turn gently several times upside down to mix. Avoid producing air bubbles, even nonvisible ones.

8. Measure fluorescence intensity in arbitrary units.

9. Repeat steps 7 and 8 until fluorescence intensity no longer changes. The fluorescence intensity is then approximately I_∞.

10. Calculate a $= I/I_\infty$ and then $(1\text{-}a)/a^2$ for each salt concentration (0, 300, 600, 900, etc. mmol/l). Plot on double logarithmic paper. Evaluate according to Eq. (7).

Results

● The Tryptophan Fluorescence Spectrum of RNP Particles

RNP particles consist of a number of "core" proteins. In the RNP particles of HeLa cells these are:
- Protein A1: molecular weight 32 000 Da
- Protein A2: molecular weight 34 000 Da
- Protein B1: molecular weight 36 000 Da
- Protein B2: molecular weight 37 000 Da
- Protein C1: molecular weight 42 000 Da
- Protein C2: molecular weight 44 000 Da

Each of the proteins consists of some 300 amino acids. Since statistically every 68th amino acid is a tryptophan, all proteins are statistically expected to reveal tryptophan fluorescence. Figure 3.1 shows the fluorescence emission spectrum af a RNP particle from HeLa cells at an excitation wavelength of 280 nm. As expected, this is clearly a tryptophan spectrum – the emission maximum is at 337 nm (Schenkel et al. 1989), indicating that tryptophan is in a hydrophobic microenvironment.

Since complete RNP particles have a few tens of tryptophans, the spectrum provides information only on those governing their fluorescence. It is also possible that a single or a few tryptophans are in a hydrophilic environment, in spite of the fact that the complete spectrum is hydrophobic. This may be a serious problem when one is interested in the function of a very specific tryptophan in the complex. However, in most cases bulk properties are of interest, for example when one asks what forces may stabilize an RNP complex. For such studies spectra of large complexes are perfectly adequate. Also, in quality control checks, a

Fig. 3.1. Fluorescence spectrum of RNP particles. The excitation wavelength is 280 nm. Part of the scattering peak at this wavelength can be seen on the *left*. The emission maximum occurs at 337 nm, typical for tryptophan in a hydrophobic environment. (Schenkel et al. 1989, with kind permission of the Biochemical Society and Portland Press, London UK)

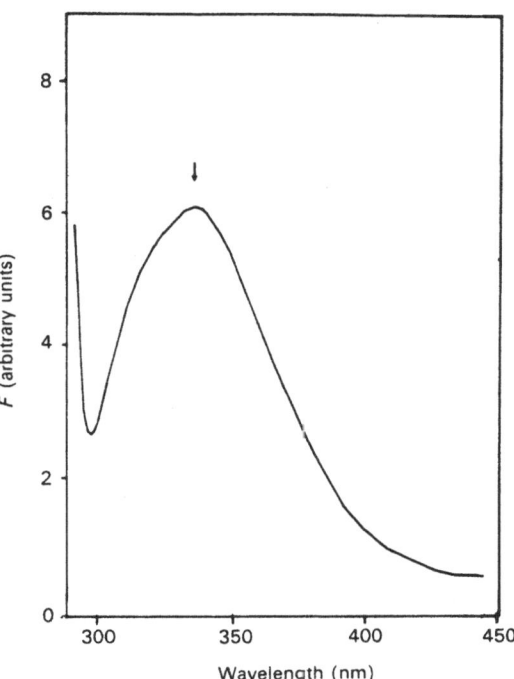

shift of this spectrum towards higher wavelengths would indicate problems in the preparation procedure. We will however see that occasionally more information can be deduced from a fluorescence spectrum when nothing changes.

- Are Tryptophans Involved in Protein-Protein and Protein-RNA Binding in RNPs?

  The fact that complete RNP particles have a fluorescence maximum at 337 nm (see above) indicates that the tryptophans are in a hydrophobic environment. This in turn suggests that they may be in the interior of the RNP particle. Additional information may be obtained from an experiment in which the particles are dissociated into their constituents. Such a dissociation can be achieved by 1 M NaCl, which obviously weakens the bonds stabilizing RNP particles. Two limiting outcomes with respect to the tryptophan fluorescence are possible:

  - The fluorescence maximum shifts to 355 nm, i.e., all tryptophans switch from an hydrophobic to an hydrophilic microenvironment. This would indicate that the tryptophans are at the surface of the individual proteins and that they can thus be directly involved in the protein-protein interactions stabilizing RNP particles.

– The fluorescence emission maximum remains at 337 nm. This would indicate that the tryptophans are buried in the single proteins and thus may not be involved in protein-protein interactions. Such a finding has, however, to be checked by solute quenching experiments (see above).

Such a salt dissociation experiment has been performed with RNP particles from HeLa cells. No change in the fluorescence emission maximum was found.

Are tryptophans involved in protein-RNA interactions? This question can now also be answered by salt dissociation. Due to the Förster energy transfer (discussed above), a tryptohan close to RNA should be dark. Upon dissociation of a protein-RNA complex it should light up, i.e., the fluorescence intensity of the solution should increase.

Figure 3.2 shows such a salt titration experiment from 0 to 2 M NaCl. No dramatic change in fluorescence intensity is observed. Thus one can conclude that no tryptophan-RNA

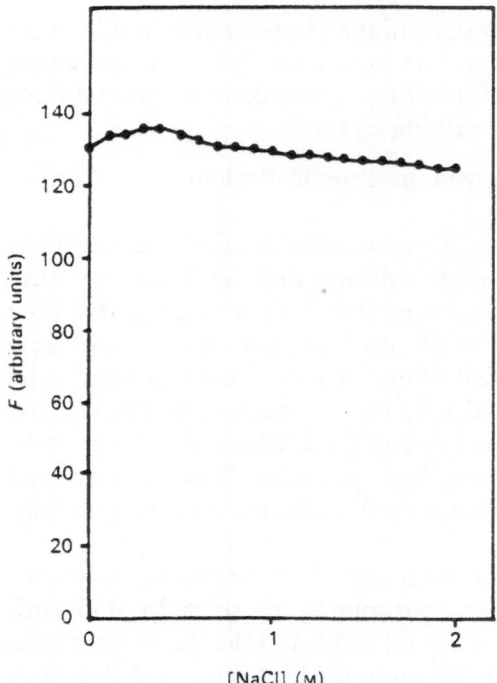

Fig. 3.2. Fluorescence intensity at the fluorescence maximum of RNP particles (337 nm) as a function of NaCl concentration. Since no change is seen, tight interactions of tryptophans with RNA can be excluded. (Schenkel et al. 1989, with kind permission of the Biochemical Society and Portland Press, London UK)

F (arbitrary units)

[NaCl] (M)

fluorescence energy transfer occurs in RNP particles from HeLa cells, i.e., tryptophan is not involved in protein-RNA interaction. In this case we cannot use energy transfer as an intramolecular ruler. An example will be discussed below for a protein-DNA complex revealing tyrosine fluorescence in which one of the tyrosines interacts with DNA; this can be seen in a salt dissociation experiment.

● Solute Quenching of Intact and Dissociated RNP Particles
In order to get a more detailed view of the role of tryptophans, intact RNP particles and RNP particles dissociated by 1 M NaCl were tested for accessibility of their tryptophans to ionic quenchers. Figure 3.3 shows the Stern-Volmer plot according to Eq. (4) for the free amino acid tryptophan, RNP from HeLa cells at low salt and at 1 M NaCl. The plots are not completely linear. This, however, is no surprise. Since probably many tryptophans in slightly different microenvironments are responsible for the fluorescence of the RNP particles, each one will have a slightly different Stern-Volmer constant, adding up to curvature of the plots. With this in mind, the plots are still remarkably linear, i.e., the microenvironments of the trypto-phans do not differ dramatically. This conclusion has been checked by the results of time-resolved fluorescence spectroscopy (data not shown).
The Stern-Volmer constants for the six experiments shown in Fig. 3.3 are listed in Table 3.3. A number of details can be obtained from these data:
– I^- is a much more efficient quencher of trytophan fluorescence than Cs^+, since with free tryptophan the Stern-Volmer constant is approximately seven times that of Cs^+.
– The relative accessibility of the tryptophans in the intact RNP particle is 13–16 % that of free tryptophan. There is obviously no difference between positive and negative quencher, the environments of the tryptophans are electro-neutral. (Note however that for a completely exposed tryp-tophan in a protein, one expects an accessiblility of only 50 and not 100 % since one half of the space is "covered" with protein.
– Upon dissociation, the tryptophans become more exposed. Now, however, there is a significant difference between positive and negative quencher. Obviously, in the intact par-ticle, some negative charge is neutralized and becomes liberated upon dissociation. It may be this charge which

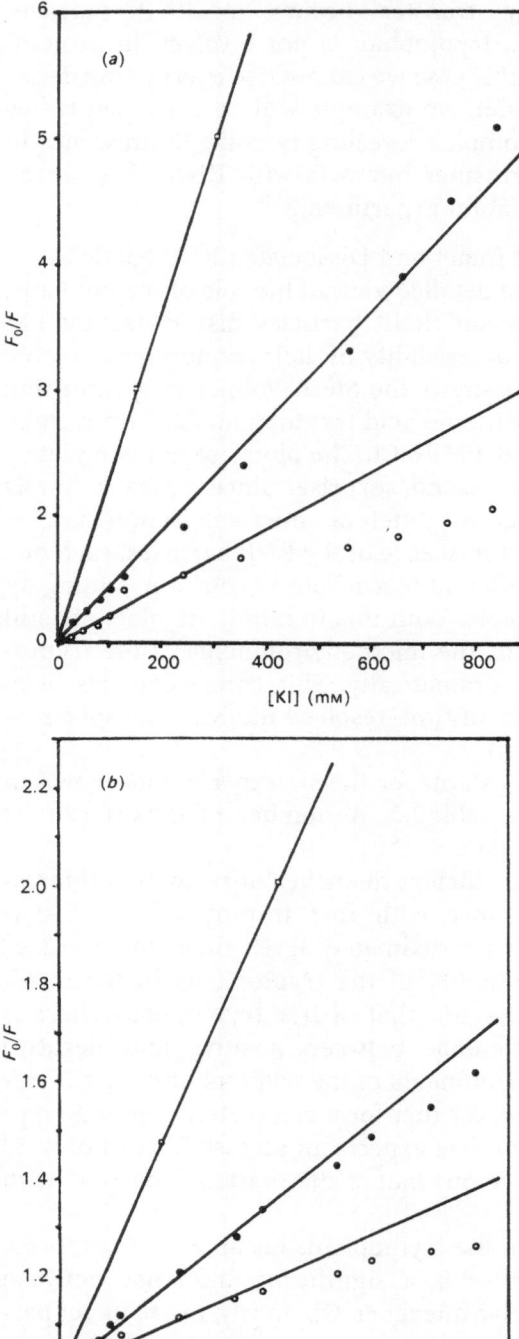

Fig. 3.3. Stern-Volmer plots for CsCl (b) and KI (a) quenching of intact RNP particles and dissociated core proteins. For comparison, the quenching of free tryptophan (steepest slope) is included. (Schenkel et al. 1989, with kind permission of the Biochemical Society and Portland Press, London UK)

Table 3.3. Stern-Volmer constants (in l/mol) for the plots of Fig. 3.3. In parentheses are the relative values, i.e., the corresponding Stern-Volmer constants divided by that of free tryptophan. These values multiplied by 100 can be regarded as the percentage of free space from which the tryptophans can be accessed

	Cs^+ quenching	I^- quenching
Free Trp	2.38	16.7
RNP, physiological	0.45 (0.16)	2.2 (0.13)
RNP, dissociated	0.83 (0.35)	4.3 (25.7)

stabilizes the RNP particle via protein-protein interactions. An electrostatic stabilization is consistent with the fact that the RNP particle can be dissociated by NaCl.

The experiments reported so far had used tryptophan fluorescence. The following two experiments are not on RNP parti cles. They show what information can be obtained when the fluorescence is governed by tyrosines.

- The Role of Tyrosines in Protein-DNA Interactions in the Filamentous Bacteriophage Pf1

The bacteriophage Pf1 (related to fd or M13) is a virus which infects coli bacteria (*E. coli*). Basically it consists of a hollow cylinder made up of several thousand identical coat protein molecules and a single-stranded DNA in the central hole of this cylinder. The distance between two bases in this DNA (the step distance in the DNA ladder) is almost twice as large as that of other known DNA structures. One reason for this large distance may be that an aromatic amino acid intercalates between the DNA bases and thereby expands the single-stranded DNA along the phage axis.

With Pf1 this problem is particularly simple since each individual coat protein molecule only has two tyrosines and no tryptophan. In fact, the whole phage shows a clear fluorescence peak at 305 nm, i.e., where tyrosine has its fluorescence maximum. In addition, it turns out that one of the tyrosines, tyrosine 25, is exposed to solvent. In principle one might quench the latter by adding a strong quencher, for example KI (see above). But there is a simple, even more efficient, method available: For labeling with iodine, kits are available which add two iodine atoms to a tyrosine provided the latter is exposed to solvent. The fluorescence of this amino acid is then totally quenched. By comparison of noniodinated with iodinated phage one can measure the fluorescence of the second tyrosine (Tyr 40). Figure 3.4 (Greulich and Wijnaendts 1984) shows the fluorescence of native and iodinated Pf1 phage.

The iodinated phage is almost dark (residual fluorescence maximal 5 % of the total phage fluorescence). This may have two reasons: (1) tyrosine 40 is also affected by iodination; (2) tyrosine 40 transfers its energy to DNA. An estimate on the distance between tyrosine 40 and the DNA bases can then be given.

The first possibility can be excluded since the coat protein is an a-helix and a distance of 25 amino acids corresponds to 40 Å. At such a distances no energy transfer from tyrosine 40 to the iodinated tyrosine 25 is expected, since the Förster distance for tyrosine-tyrosine energy transfer is of the order of 15 Å and soon vanishes beyond that distance. Thus tyrosine 40 obviously transfers more than 95 % of its fluorescence energy to the DNA. From Eq. (5) one can then estimate that the tyrosine-DNA distance is considerably less than 7 Å. Therefore, the fluorescence data are consistent with the assumption that tyrosine 40 interacts with the DNA.

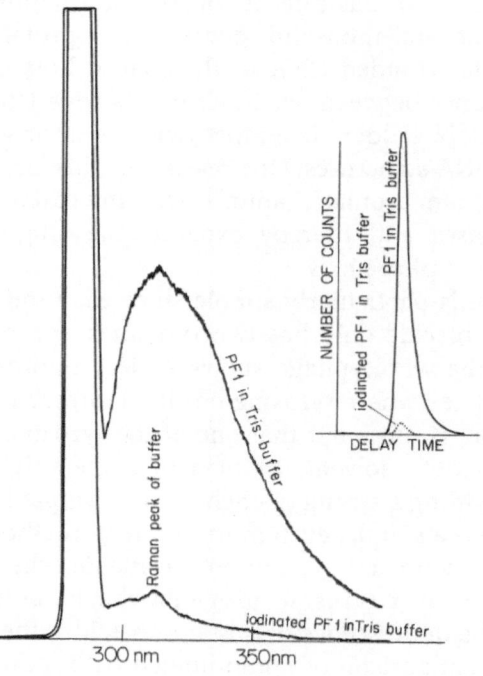

Fig. 3.4. Fluoresence spectra of native Pf1 and of phage with iodinated (i.e., quenched) tyrosine 25. Both have similar scattering peaks at 280 nm, indicating that the concentrations of Pf1 in both experiments are similar. The iodinated phage reveals only marginal tyrosine fluorescence. The steady state fluorescence data are confirmed by time-resolved photon counting (*insert*). (Greulich and Wijnaendts van Resandt, with kind permission of Elsevier Science-NL, Amsterdam)

- The role of Electrostatic Interactions in Protamine-DNA interactions

One particularly interesting example of studies of electrostatic interactions is the protamine thynnine. It consists of 31 amino acids, 21 of which are positively charged arginines. In the center of its sequence a single tyrosine serves as a natural reporter group for fluorescence studies. This molecule can be used to test the electrostatic behaviour of arginines. From the other positively charged amino acid, lysine, it is known that lysine develops a full electrostatic bond with DNA. How does arginine behave? When thynnine is bound to DNA, the tyrosine transfers its energy totally to the DNA, i.e., the thynnine-DNA complex is dark. By adding NaCl, electrostatic interactions are shielded, thynnine dissociates from the DNA and fluorescence occurs, since according to Eq. (5) fluorescence energy transfer is no longer possible at the now large tyrosine DNA distance. Figure 3.5a shows the increase in fluorescence upon addition of NaCl, Fig. 3.5b the plot of the data according Eq. (7).

The result is indeed approximately a straight line. From the slope one can calculate that the 21 arginines form only four full electrostatic bonds. This, and the fact the the intersect of the line with the vertical axis is not at zero, is consistent with

Fig. 3.5 a Salt dissociation curve of a thynnine-DNA complex. **b** Plot of these data according to Eq. (7) (see text). (Ausio et al. 1984, reprinted by permission of John Wiley & Sons, Inc.)

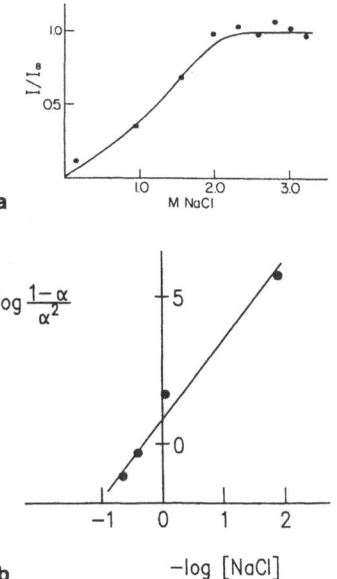

the fact that arginine, unlike lysine, does not develop a full electrostatic bond with DNA. With a similar molecule containing lysines instead of arginines one would have expected a line with a much steeper slope but with a intersect closer to zero.

For experiments of this type it is irrelevant if the reporter molecule is tyrosine or tryptophan. Also, when only part of the aromatic amino acids are close to the DNA, the experiment can still be performed. In that case, the fluorescence at zero salt concentration in Fig. 3.5b is not zero. Then, instead of I_∞, the term $(I_\infty - I_0)$ has to be used, where I_0 is the residual fluorescence at zero salt concentration.

Troubleshooting

- Absorption not reproducible.
  - Cuvette may be the reason: Wash and dry cuvette. Measure OD of empty cuvette at 280 nm. If washing has resulted in improvement, repeat until no further improvement can be achieved. If still residual fluorescence is seen cuvette has to be immersed overnight in sulfur-hydrochloric acid and then rinsed with distilled water at least 10 times.
  - Scattering peak too high: Degas all solvents and stock solutions. If this does not help, prepare new ones, even if it means several hours of work.

- Fluorescence not reproducible.
  - If reasons above are already excluded: Check if fluorescence excitation lamp in the spectrometer is working properly. Consider that the typical working time of such a lamp may be only 4–5 months in a lab where the spectrometer runs all day. Also, frequent switching on and off may reduce the lifetime.

- Extremely large light scattering peak where it is not expected.
  - When the light scattering peak at the excitation wavelength becomes larger than typical for this type of preparation, this probably indicates the formation of molecular agglomerates, i.e., your sample may be denatured.

References

Ausio J, Greulich KO, Haas E, Wachtel EJ (1984) Characterization of the fluorescence of the protaminee thynnine and studies of its binding to double stranded DNA. Biopolymers 23:2559–2571

Eftink M, Ghiron CA (1981) Fluorescence quenching studies with proteins. Anal. Biochem 114:199–227

Greulich KO, Wijnaendts van Resandt R W (1984) Estimation of Tyr40-DNA distance in the filamentous phage Pf1 by analysis of its intrinsic fluorescence properties. Biochim Biophys Acta 782:446–449

Schenkel J, Appel I, Schwarzwald R, Bautz EKF, Wolfrum J, Greulich KO (1989) Fluorescence studies on the role of tryptophan in heterogeneous nuclear ribonucleoprotein particles of HeLa cells. Biochem J 263:279–283

Schwarzwald R, Greulich KO (1988) Tyrosine fluorescence energy transfer as a probe for interactions of proteins with DNA. Ber Bunsenges Phys Chem 92:447–450

Stryer L (1978) Fluorescence energy transfer as a spectroscopic ruler. Ann Rev Biochem 47:819–846

Szmacinski H, Wiczk, Fishman MN, Eis PS, Lakowicz J (1996) Distance distribution from tyrosyl to disulfide residues in the oxytocin and [Arg^8] vasopressin measured using frequency domain fluorescence resonance energy transfer. Eur Bioph J 24:185–193

Zhang L, Hermans J (1996) Hydrophilicity of cavities in proteins. Proteins 24:433–438

Suppliers

- Fluorescence spectrometers (prices starting from $ 17 000), in alphabetical order: Anthos, Hitachi, Kontron, Perkin Elmer, Polytec, SCM Aminco, Shimadzu
- Quartz cuvettes ($ 100–130): Hellma (QS quaility)
- Solvents and reagents: (Uvasol Solvents) Merck; Iodination kit BioRad

Most other material can be purchased from any lab supplier.

Chapter 4

Procedures for Three-Dimensional Reconstruction from Thin Sections with Electron Tomography

Ulf Skoglund*, Lars-Göran Öfverstedt,[1]
and Bertil Daneholt

Introduction

Electron tomography

Electron tomography (ET) is a method for three-dimensional (3-D) reconstruction of single, transparent objects from a series of images (i.e., a tilt series) recorded with a transmission electron microscope (EM) (Fig. 4.1). The method is related to the procedures used in medical tomography. The 3-D reconstructions are usually computed from the digitized tilt series after a radial weighting scheme has been applied to the Fourier-transformed data. The ET method can be used to reconstruct in 3-D any object that is transparent enough for projection imaging with the transmission EM. This means that specimens of biological origin are usually available for ET 3-D reconstruction, whereas, e.g., colloidal gold particles are not. An ET-calculated 3-D map could be based on projections from objects that have been visualized by EM in several different ways, i.e., from stained or unstained objects, objects visualized at different energy loss levels or objects embedded in different media. The combined analysis of the 3-D structure, imaged in different ways, could thus become very informative. The general applicability also means that it is not restricted to symmetrical or regularly arranged objects, nor to objects with preferred orientations on a support grid. In its present state, the ET method allows reproducible 3-D reconstructions of single molecular objects, with a resolution in the range of 5 nm, of complex cellular specimens. For isolated objects, free of interfering cellular substances, a somewhat higher resolution can be achieved. The ET method

* Corresponding author: Ulf Skoglund: Tel.: (+46)-8–7287364;
  Fax: (+46)-8–313529; e-mail: ulf.skoglund@cmb.ki.se
[1] Laboratory of Molecular Genetics,
  Department of Cell and Molecular Biology,
  Medical Nobel Institute, Karolinska Institute, 17177 Stockholm, Sweden

covers the intermediate resolution range where there is no other physical technique available to analyze single molecular complexes.

It is becoming increasingly clear that many molecules carry out their function in multimolecular complexes and that our knowledge about these complexes is still fragmentary, relying to a large extent on the degree of preservation during biochemical isolation. For example, apart from well known multimolecular structures such as the ribosome and the spliceosome, transcription and replication machineries are structures which become increasingly complex as additional subcomponents are identified. The ET method makes it possible to study in 3-D these supramolecular assemblies, as well as particles, membranes and filaments.

In the following we will outline the implementation of the method in our laboratory and the practical way to carry out an investigation. We have used the ET method to analyze specimens of various origins (e.g., ribosomes in *E. coli*, Balbiani ring (BR) pre-messenger ribonucleoprotein (pre-mRNP) particles, HIV in infected cells, and myelin sheets), but in describing the ET procedure we will show how the analysis proceeds for one specimen, the BR pre-mRNP particles.

Biological system

Pre-mRNAs, synthesized by RNA polymerase II in a eukaryotic nucleus, associate with proteins to form RNP complexes (Dreyfuss et al. 1993). Electron micrographs of growing RNA on an active gene indicate that RNP formation starts immediately when the growing RNA chain leaves the polymerase (Miller and Bakken 1972; Daneholt 1992). These RNPs consist mainly of a complex between RNA and the hnRNP proteins, a class of general RNA-binding proteins (Dreyfuss et al. 1993). A few major hnRNP proteins dominate the protein part of the RNP complex (Chung and Wooley 1986). RNP particles have a sequence-specific distribution of RNA-binding proteins along their sequence (Dreyfuss et al. 1993).

Pre-messenger RNA is usually modified through a series of reactions (capping, splicing and polyadenylation) during synthesis and transport out to the cytoplasm (Darnell 1982). Very little is known about the transport mechanisms for pre-mRNP, but it has been shown that transport to the cytoplasm goes through the nuclear pores (Stevens and Swift 1966; Dworetzky and Feldherr 1988; Mehlin et al. 1992). A prerequisite for pore translocation seems to be that the splicing and polyadenylation reactions

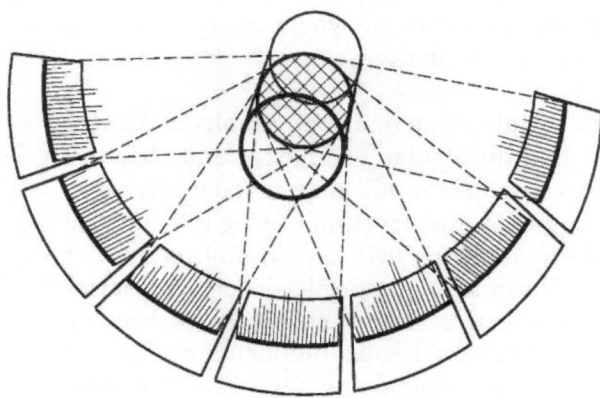

Fig. 4.1. The tomographic principle. The three-dimensional (3-D) reconstruction is performed as a stack of 2-D reconstructions, each being performed by R-weighted back projection from lines in a set of aligned electron micrographs

have been completed. Only fully processed mRNA seems to be released to the cytoplasm.

To understand phenomena such as RNP formation, transport, modification (e.g., splicing) and degradation, it is important to monitor a specific pre-messenger RNP particle in the electron microscope during the various processing steps. The BR pre-mRNP particle offers such a possibility.

Outline

The BR pre-mRNP particle has been studied with ET mainly as it appears in situ. Conventional EM techniques have been used for preparation of the material. The steps sufficient for an ET 3-D single particle reconstruction from a thin section are summarized in Fig. 4.2. Steps 3–6 involves computer processing and the use of the implemented set of programs. Currently our processing tools consist of 71 different computer programs (with some more under development). Each processing program is run from a unix shell-script (a "program.sh" file), in which parameters and file names are entered, and generates besides processed data files a general output file (a "program.out" file) which monitors the result of the computation. We also use some computer graphics programs to visualize images and 3-D reconstructions. Two of these multifunctional graphics programs are heavily used

Fig. 4.2. The principal steps encountered in electron tomography

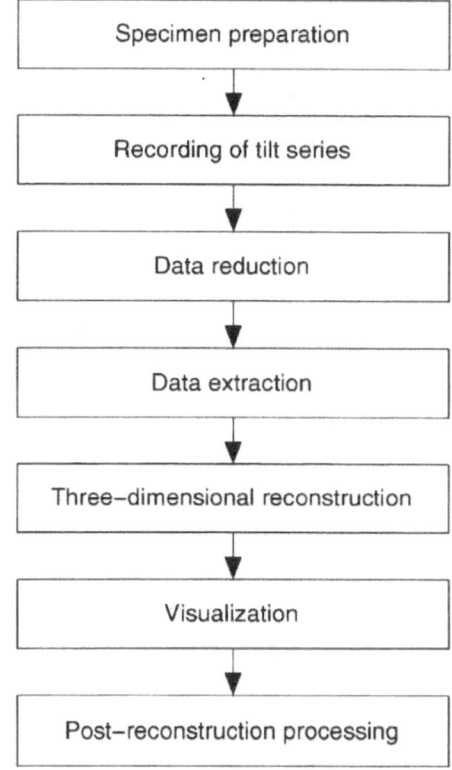

at various stages of the data processing. A large part of the "tool-set" of computer programs, however, are not used in all the different ET applications. In fact, only a limited set of processing tools have to be understood in order for a beginner to succeed with a properly carried out ET reconstruction.

Materials

Reagents

- Embedding medium: agar resin 100 (Agar Aids Ltd., Stansted, UK)

Equipment

- Electric oven with thermostat set at 60 °C
- Ultramicrotome (LKB 2088, Ultrotome V; Cambridge Instruments Ltd., Cambridge, UK)
- Transmission electron microscope (Zeiss CEM 902) equipped with a ± 60° goniometer and a liquid nitrogen cold trap

- Rotating drum scanner (Optronics P-1000) run by a PC (IBM 286 compatible) that is connected to the main computer via a network file system (NFS)
- Main computer (DEC 3000 Model 400 Alpha Workstation, Digital Equipment Corporation) under Digital UNIX version 3.2C
- Graphic display (Crimson, Silicon Graphics, Inc.) under IRIX version 5.3

Buffers – 0.05 M sodium cacodylate-HCl buffer, pH 7.2

Procedure

ET implementation at the Karolinska Institute

Use of the ET method since 1983 has resulted in the collection and development of a full system of computer software and technical procedures which have been successfully applied in a wide variety of projects. The design of the programs has been such that while users can make full use of their potential, they nonetheless retain a maximum of flexibility, allowing for further modifications should the need arise. Thus the software consists of a series of programs rather than one large program with several internal options. This set-up has the disadvantage of being slightly more difficult to learn, but offers the advantage of flexibility and the ability to improvise during data processing. Most of the applications of the ET procedure have been to reconstructions of stained and plastic-embedded biological specimen from thin sections (Fig. 4.3). Consequently, most of the software has been developed accordingly (Fig. 4.4).

Specimen Preparation

1. Isolate salivary glands from *Chironomus tentans* fourth instar larvae.

2. Fix the glands in 2% glutaraldehyde in 0.05 M sodium cacodylate-HCl buffer, pH 7.2, at 4 °C for 2 h and rinse four times 15 min with the buffer. Transfer the glands to 1% osmium tetroxide in the buffer for 1 h at 4 °C and rinse as above.

3. Dehydrate the glands at room temperature in a graded series of ethanol (from 70 to 100%).

Fig. 4.3. The procedure
for specimen preparation
for electron tomography

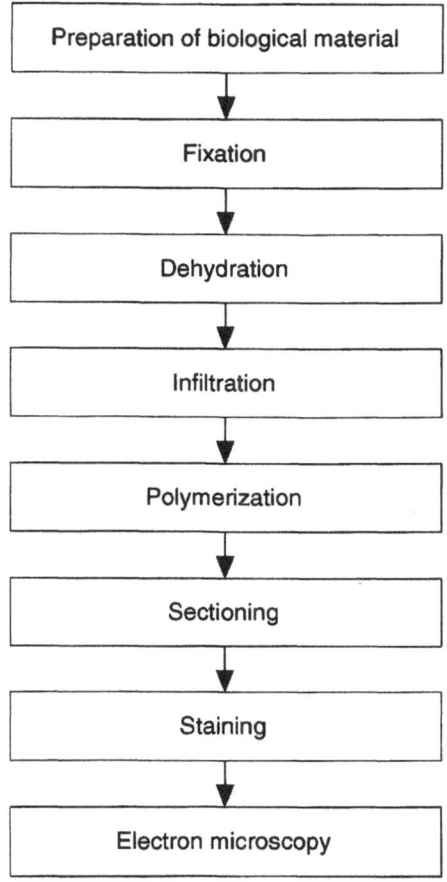

4. Infiltrate the glands with agar resin 100; the agar concentration is increased from 25, to 35, and finally to 50 % over a period of 60–90 min. Keep the glands at 50 % overnight and embed in 100 % agar (in gelatin capsules) for 3×2 h.

5. Let the resin polymerize at 45 °C for at least 2 days and at 60 °C for at least 3 days.

6. Section the glands in an ultramicrotome (e.g., LKB 2088). To locate optimal Balbiani rings: make thick sections, stain with toluidine blue, and study the sections in a light microscope. Subsequently, collect thin sections on Formvar-coated, single-slot grids. The interference color of the sections should be silver, corresponding to a thickness of 60–90 nm.

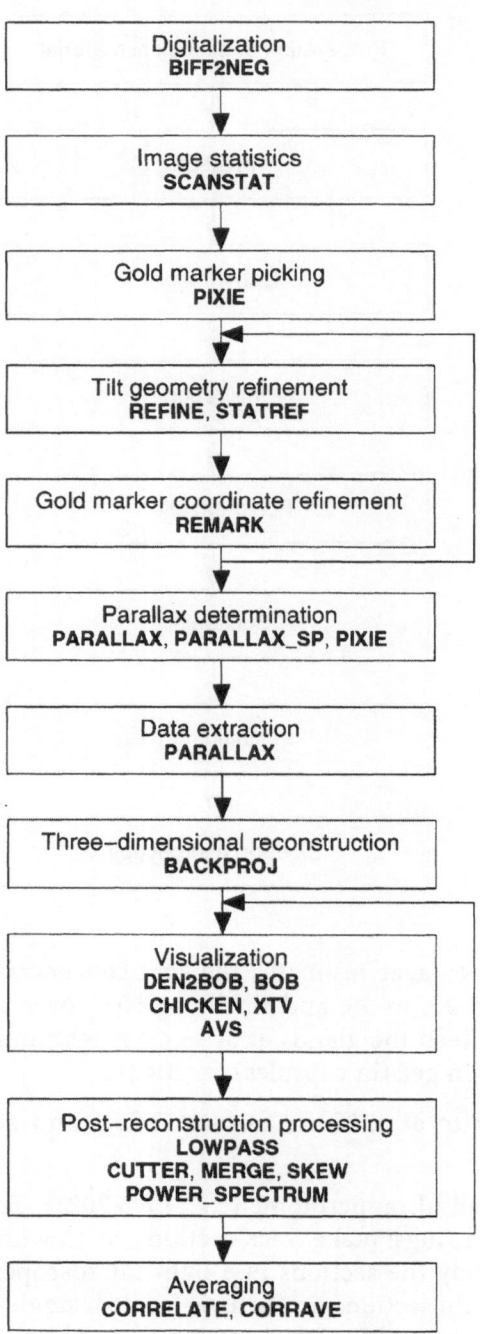

Fig. 4.4. Flow chart of the computation procedure for three-dimensional reconstruction by electron tomography

7. Stain the specimens for 5 min at room temperature in a saturated solution of uranyl acetate in 50 % ethanol. Rinse briefly in 50 % ethanol and thoroughly in distilled water, and then dry. Stain at room temperature for 2 min in 0.4 % lead citrate at high pH and in the dark. Rinse briefly in 0.02 M NaOH and thoroughly in distilled water, and then dry.

8. Put a droplet of 10 nm colloidal gold particles on top of the sections and remove the surplus liquid after 1 min.

Recording of a Tilt Series

1. The specimen is entered into the EM (Zeiss CEM902) and a suitable area containing one or more objects is selected. The area should contain a sufficient number of gold markers (10–15) for the subsequent geometry alignment of the tilt series.

2. Record a pre-micrograph at $0°$ tilt angle.

3. Record the tilt series. Start at one extreme (e.g., +60°) of the goniometer and record images, e.g., at 5° intervals, to the other extreme (Fig. 4.5). Refocusing and astigmatism correction between each image should preferably be done at an area adjacent (along the tilt axis) to the recorded area in order to minimize electron beam damage.

4. Record a post-micrograph at $0°$ tilt angle.

5. Develop EM negatives.

6. Compare pre- and post-micrographs for evident beam damage. If these micrographs differ significantly, the tilt series must be rejected and a new one recorded.

Data Reduction

1. Check all micrographs in the tilt series for quality (defocus and astigmatism). **Note:** The quality of the 3-D reconstruction is heavily dependent upon the quality of the data (micrographs).

2. Scan all EM negatives in the tilt series in the drum scanner run by a program that is implemented on a local PC (IBM 286

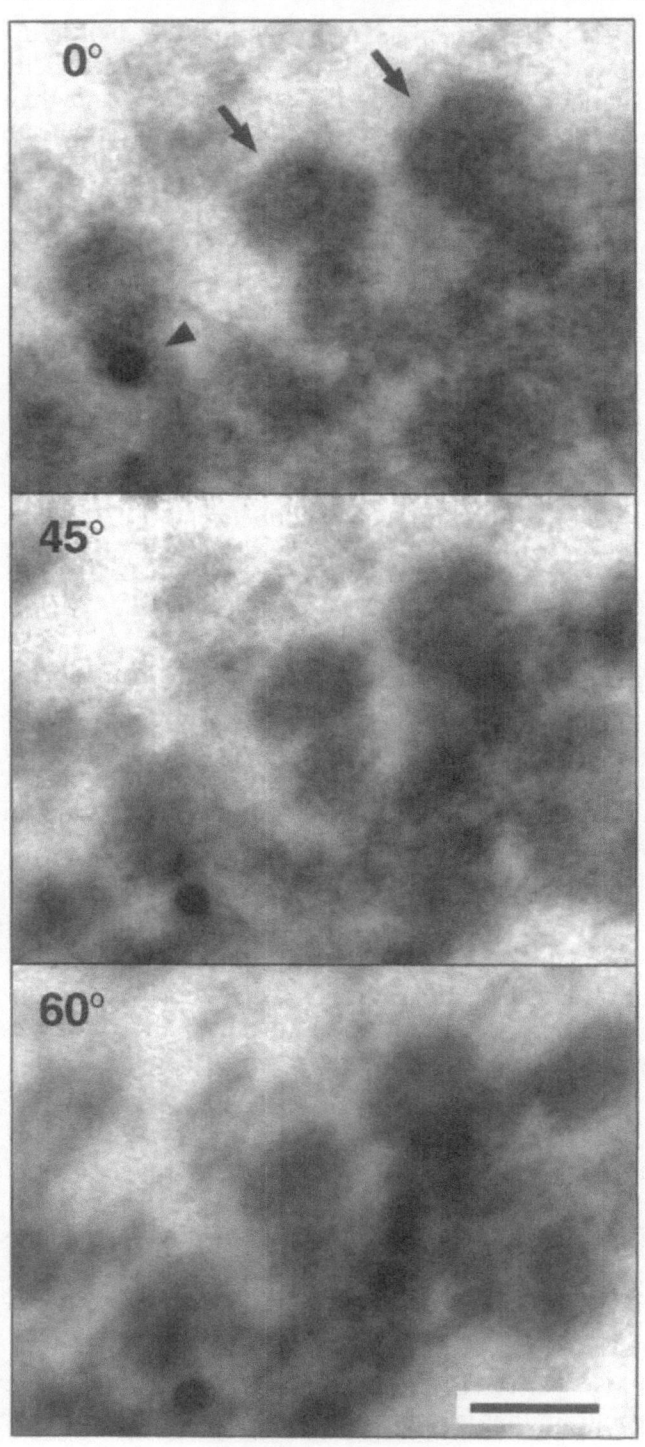

◀ **Fig. 4.5.** Electron micrographs of a segment of an active Balbiani ring (BR) gene in a cell nucleus of an embedded and sectioned salivary gland of *Chironomus tentans*. The stalked granules represent growing RNP particles; two of these have been pointed out with *arrows*. An added gold particle (*arrowhead*) can also be seen. Three images (out of 25 in a tilt series) recorded at 0°, 45° and 60° tilt angle are shown. *Scale bar* is 500 Å

---

compatible). The PC program dumps the scanned data via NFS to a directory on the main computer as separate files with the extension ".bif". These "*.bif" files are converted by the program BIFF2NEG into a format used by several of the subsequent processing and graphical display programs. The converted files have the extension ".neg". To check if a negative is properly scanned, you view the image on the graphical display with the PIXIE program. If you want to plot the picture on a laser printer instead, you could run the TONER program, which creates a PostScript printout.

3. After scanning, the SCANSTAT program has to be run on the "*.neg". This program puts some statistics parameters in the first block of the "*.neg" files. **Note:** It is necessary to run SCANSTAT on **all** "*.neg" files to avoid trouble with wrong coordinates in the subsequent processing.

4. Denote an order number for those gold markers that are clearly visible in all scanned tilts. For each "*.neg" file, use the **pick gold** command in the **gold** menu in the PIXIE program to pick the coordinates of the gold markers, in order, into a corresponding "*.au" file. Use the **test gold** command to check that coordinates are properly centred and to create a gold marker mini-image set for subsequent coordinate refinement. Measuring the average diameter of the gold markers and setting the box size in **pick gold** accordingly before picking the coordinates is recommended. **Note:** It is very important that the gold markers are picked in the same order throughout the tilt series.

5. Run the REFINE program, using the "*.au" files, to determine the orientation of the different pictures in relation to each other. The resulting geometry description is saved in the "refine.dat" file. Type the "refine.out" file and look at the residual values to see if a proper result has been achieved. By experimenting with the start values in REFINE, e.g., estimated

tilt angles and direction of tilt axis, it is usually possible to find a good optimum. Run the STATREF program to evaluate the accuracy of the origin definition. The average error is given in both pixels and angstroms. Try to get the origin definition as accurate as possible; an average error lower than 1 pixel for all tilts should be attained.

5a. Change the box size in the refinement of the gold marker coordinates. The selected box size in PIXIE may not be the optimal choice. Run the REMARK program with other values of the size parameter to generate new sets of "*.au" files. REMARK utilizes the set of mini-images created by the test gold command in PIXIE (see step 4). Repeat the REFINE and STATREF runs and select the set of "*.au" files that gives the best result (lowest error). **Note:** The box size in PIXIE, which corresponds to the side length of a search square, is not comparable to the size parameter in REMARK, which is the diameter of a circular mask.

5b. Exclude the gold markers with the worst origin definition as seen in "statref.out". Gold markers that are out of focus in high tilts, that are not perfectly spherical, or for some other reason give a poor origin definition can be excluded in REFINE. Run STATREF to monitor the result.

Data Extraction

The instructions provided thus far have referred to all objects in the tilt series. From now on you will only work with the object to be reconstructed.

1. Determine the x-, y- and z-coordinates for the object to be reconstructed. This will be the mid-point of the reconstructed volume. The x- and y-coordinates can easily be picked using PIXIE on the zero tilt. The parallax value (z-coordinate) for the object is determined by a limited data extraction from three images with the PARALLAX program. Use the highest tilt angles and the zero tilt ($-60°$, $0°$, $+60°$) to get the most accurate determination. Start by letting the z-coordinate be 0.0 (i.e., average z-coordinate for the gold markers). After the PARALLAX run, you can use the PARALLAX-SP program to generate one image ("parallax.neg" file) of the three extracted areas. Look at this image with the PIXIE pro-

gram and use the **parallax** command to determine the objects parallax (deviation from extract mid-point) and adjust the z-coordinate accordingly.

2. Now, given the coordinates of the object, use PARALLAX to make a full data extraction from all tilts in the series. Calculate the extract size (in pixels) needed to fully include the object. The extracts are cut out parallax-adjusted and parallel to the tilt axis according to the parameters in "refine.dat". Run PARALLAX-SP and PIXIE programs to examine the result. If the z-coordinate is correctly determined the object should remain in the middle of all extracts. PARALLAX also permits data reduction by binning, i.e., the micrographs are scanned with a smaller pixel size which are averaged into the regular pixel size during extraction. This is highly recommended to reduce the noise level. All extracted areas are stored together in one sorted data file "extracts.srt" (Fig. 4.6) which will be input to the back-projection reconstruction program.

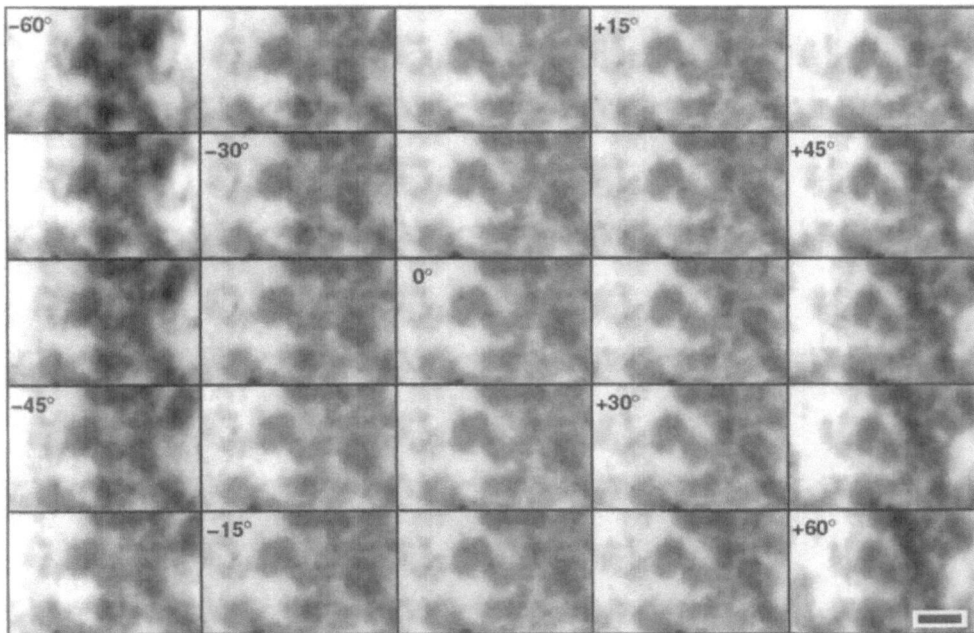

Fig. 4.6. Extracted corresponding regions from the 25 digitized electron micrographs in the tilt series shown in Fig. 4.5. The extracts are aligned such that the tilt axis is vertical and stored in a format suitable for the subsequent 3-D reconstruction. *Scale bar* is 500 Å

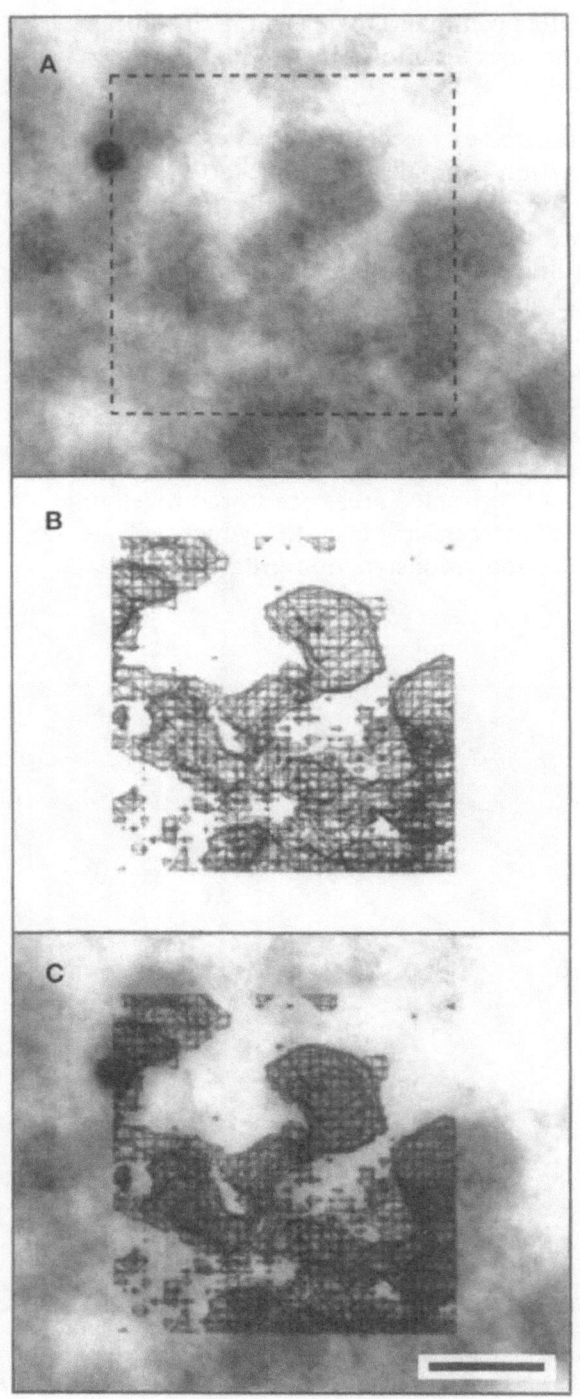

◀ **Fig. 4.7.A** Electron micrograph (0° tilt) with the position of the reconstruction indicated by the *inserted frame*. **B** Isodensity contour representation of the 3-D reconstruction shown in the 0° tilt orientation. The contours are at a threshold 1.2 root mean-square units above the average density in the 3-D volume. **C** The contoured 3-D reconstruction overlayered on the micrograph showing the relation between this contouring threshold and the gray scale of the micrograph. *Scale bar* is 500 Å

When you are satisfied with the PARALLAX output, you have extracted all data needed to run the back-projection program. The space-consuming "*.neg" files can be stored on tape and then deleted from the disk, since they will not be used any more.

Three-Dimensional Reconstruction

1. Perform the filtered back projection with the BACKPROJ program using the sorted extracts ("extracts.srt") from all tilt angles as input. Be sure to use the same parameter values for section thickness, etc., as were used in the PARALLAX run. The information in all extracts is back projected to form the 3-D volume; first a 1-D slice from each projection is Fourier-transformed, then multiplied by the radial weighting scheme, Fourier-transformed back and then summed for the points (x, y) in a 2-D slice. This is performed "slice by slice" in the extracts to make all the necessary 2-D slices, which then, put together, stands for the final 3-D volume.
If you want to add a few slices to a reconstruction made previously, you can reconstruct the extra slices with BACKPROJ and add them to the old reconstruction with the MERGE program, provided that the new slices were already included in the parallax extracts. (However, you should be aware that the density scaling parameters could be off unless you supply your own set of scale factors – an option in BACKPROJ.)
To visualize single sections of the reconstruction, you can either make plots by running TONER and then printing them on the laser printer, or make "neg-files" of them and visualize them on the SGI-screen by PIXIE. The information in the single sections can give you a feel for how well you have estimated the resolution for the BACKPROJ input. The reconstructed density can be compared to the original projection

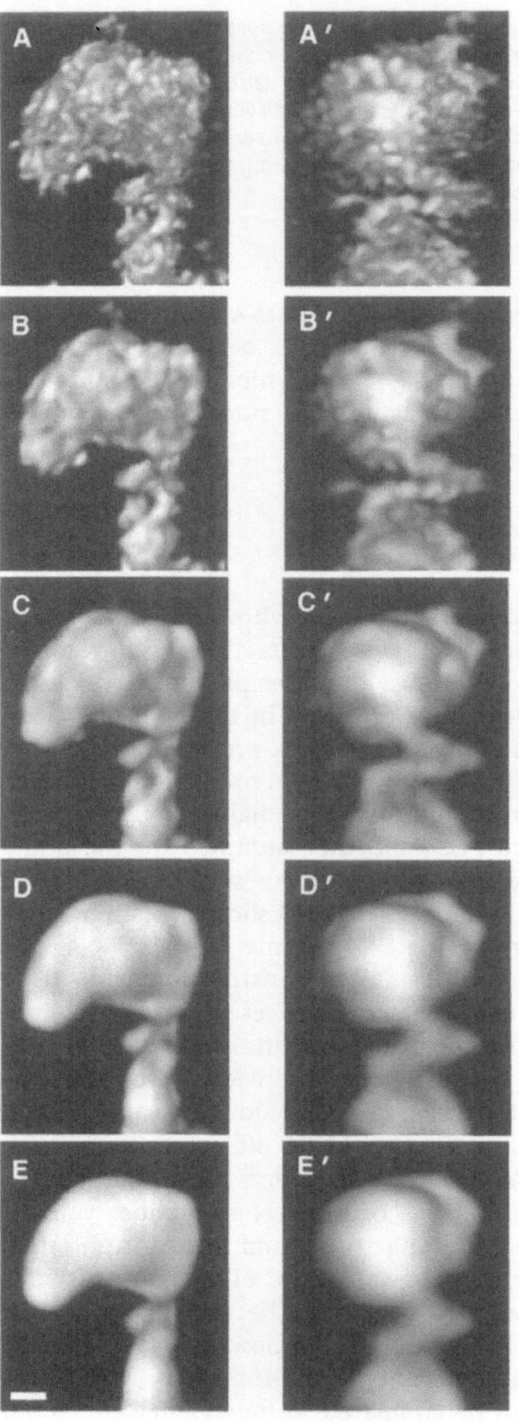

Fig. 4.8A–E. The influence of low pass filtering of the reconstruction. A cut-out central region of the 3-D recontruction containing the "head" of the growing pre-mRNP particle is visualized after low pass filtering to different limiting resolutions. The structure is shown as volume-rendered models in the 0° orientation (see Fig. 4.7) and after a 90° rotation around a vertical axis (denoted by **A'–E'**). The limiting resolutions are 15 Å (**A**), 30 Å (**B**), 45 Å (**C**), 60 Å (**D**) and 75 Å (**E**). The optical resolution, calculated from the number of tilts and specimen thickness, is approximately 45 Å. *Scale bar* is 200 Å

data to check whether the parameters used in the reconstruction process were reasonable (Fig. 4.7).

2. The relative density scale factors between the tilts are calculated during reconstruction. Checking them in "back-proj.out", to see that they fall within reasonable limits, is recommended. If, for some reason, some of them deviate considerably from 1 the corresponding tilts will have a seriously wrong weight in the calculated 3-D reconstruction.

3. Use the LOWPASS program to calculate an accurate lowpass-filtered density at the desired resolution. Usually the 3-D density you get from the filtered back projection is calculated at the resolution expected from the basic resolution formula: resolution = specimen diameter$\times\pi/$ number of tilts over 180°.
   If your tilting only spans about 120° ($-60°$ – $+60°$), the resolution will be around 50 % lower in density in the beam (z) direction. The density will also appear quite edgy, elongated in the z-direction (beam direction), and noncontinuous. The simple remedy to this is to reduce the resolution of the density – as a starting point by 50 %. This might not be enough, however, so lower it more if needed. The density should look reasonably smooth and continuous when accepted (Fig. 4.8). Locally the actual resolution might be higher than expected. This happens when the thickness of the sample, given any optical contrast, is less than the average expected thickness. Make sure that this is really true before you accept a higher resolution filtering.

Visualization of the Reconstruction

The reconstruction can be illustrated in several ways. If you want to build a physical model, e.g., using balsawood, you need to print sections through the density. The TONER program does this by making a plot of the sections for a laser printer. Interactive stereo viewing and numerous manipulations of isocontours of the density can be done with the graphical XTV program which runs on a SGI computer. By varying the density iso-contouring level (Fig. 4.9), a meaningful contouring level can be established at which the appearance of the structure is consistent with the resolution as well as the fit to the projection data (Figs. 4.6, 4.7). The contours are generated by the CHICKEN program. Before you

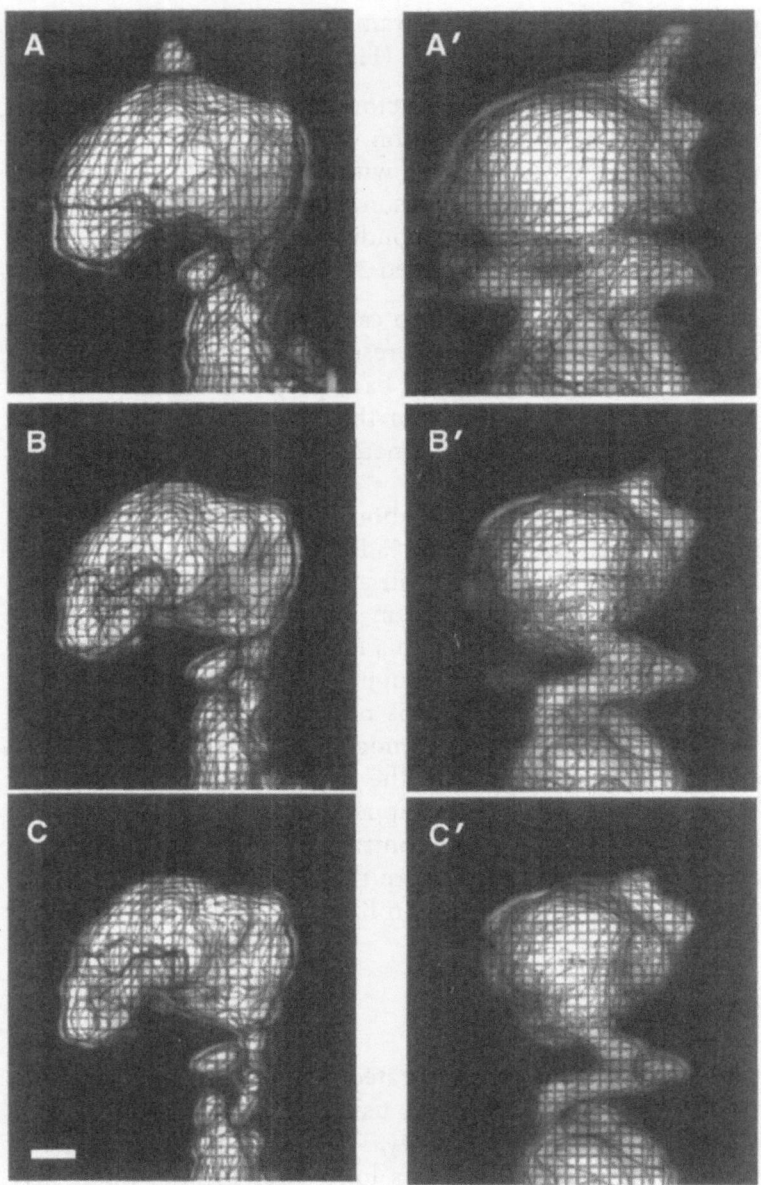

Fig. 4.9A-C. The influence of contouring level on the appearance of the structure. The central region of the 3-D recontruction is visualized at different contouring thresholds as volume-rendered models with overlayered chicken-wire representation. The orientations are as in Fig. 4.8. Contouring thresholds are 0.8 (**A**), 1.3 (**B**) and 1.55 (**C**) root mean-square units above the average density in the volume. A meaningful level for analysis of the structure can be deduced by comparison to the original micrographs (see Fig. 4.7C). *Scale bar* is 200 Å

can visualize your density, you must determine some statistical parameters of the 3-D density with GRAVITY.

You can also examine your 3-D density as a volume-rendered object in an interactive stereo-viewing mode with the BOB program on a SGI computer. Volume rendering visualizes in a convenient way the total density distribution and its variations. Usually the appearance of a volume rendered object is fuzzy near surfaces, and thus the exact surface curvatures at a particular volume of the object are difficult to assess. At an iso-contour the surface can be exactly visualized, either as iso-contour lines (XTV) or as a continuous iso-surface (XTV) (Fig. 4.10)

Post-Reconstruction Processing

1. The 3-D density can be manipulated with the CUTTER program. With this program you can change axis order and cut out any boxed region from the original density. This might prove handy, e.g., if you have several similar, randomly distributed structures in the same reconstruction that you want to compare with each other.

   Cut-outs

2. With the SKEW program you can quickly rotate and resample your density into any chosen orientation. **Note:** Points in the corners of the 3-D density may be lost during this resampling.

   Skewing

3. With the POWER-SPECTRUM program you can analyze the distribution of power in the Fourier space and plot it as a graph with the power as a function of the resolution. If you have two related structures these can be compared and their similarities compared quantitatively in the Fourier space with the SHELL-CORR program. A good, and internationally accepted measure of similarity between two structures, the phase residual, is calculated and plotted in resolution shells by the SHELL-CORR program.

   Power spectrum

4. With the density correlation programs (dcp) you can compare two densities, a probe from one density with another density to be aligned. The CORRELATE program calculates the correlation coefficient between the probe and the other density at a given orientation and also calculates in a least-squares sense the orientational and translational shifts necessary to increase the correlation between the densities. Thus it

   Averaging

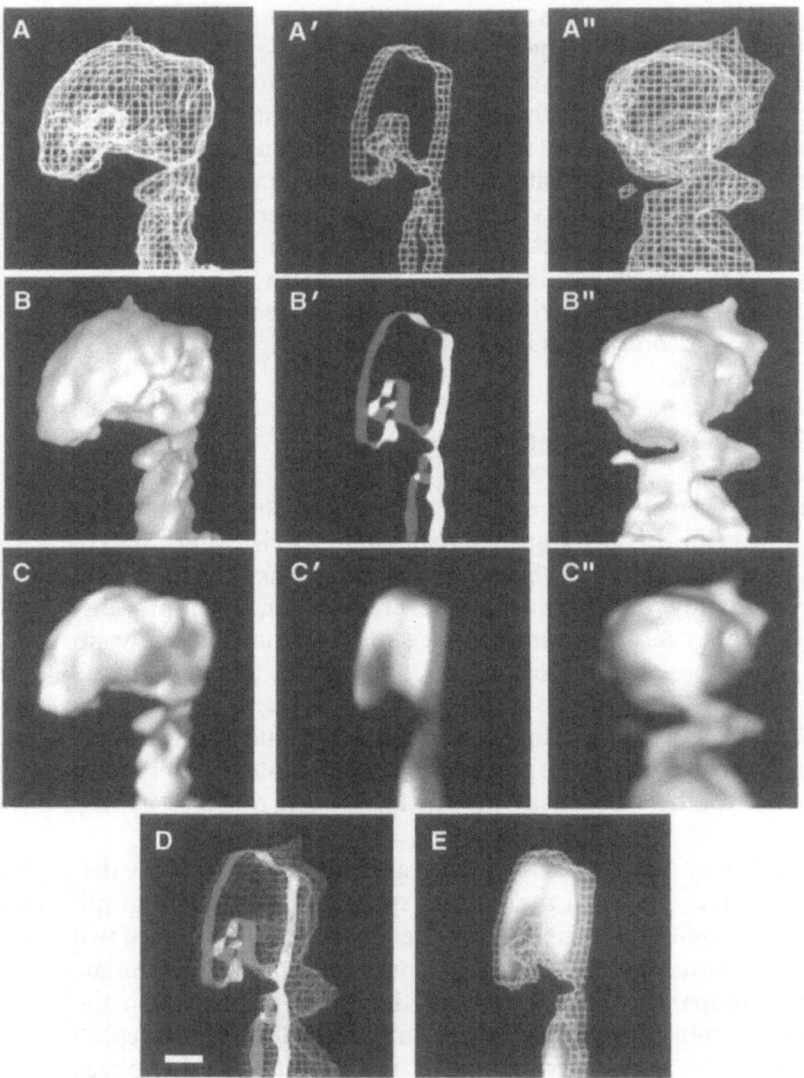

Fig. 4.10A–E. Different visualization techniques. The central region of the 3-D reconstruction is modeled by iso-density contouring (**A**), surface rendering (**B**) or volume rendering (**C**). The structure is shown in the 0° orientation and after 60°(') and 90° (") rotations around a vertical axis. The clip planes used in **A'**, **B'** and **C'** show that only volume rendering gives a full representation of the interior structure. Combinations of **A'**, **B'** and **C'** are shown in **D** and **E**. All models are at 1.3 root mean-square units above the average density. *Scale bar* is 200 Å

is possible to find the best fit between the two densities by an iterative application of the correlation procedure.

5. If you have several densities you want to compare you can run the CORRAVE program. It will optimally align the densities to each other in an unbiased way and calculate an average density at the end. All density correlations are made within a predetermined envelope (mask) that outlines the features of interest in the density.

   The average between two or more presumably similar, but independently reconstructed, densities has an increased signal/noise. If an average is desired, then the first thing to do is to find out how the independent densities are spatially related, i.e., the rotational and translational parameters that relate the densities must be known. There are several ways to establish these parameters, and the dcp described above is only one possibility. The advantage with the dcp is that the approach is completely general. No assumption is made on the existence of symmetry or due to the fact that the different copies exist in the same reconstruction. A substantial number of biological specimens do have internal symmetry, such as dimers or multimers or even the icosahedral viruses. In these cases the rotational and translational parameters can be found by approaches similar to molecular replacement techniques used in crystallography.

Results

- Formation and Transport of a Specific Pre-messenger RNP Particle

  The BR system of *Chironomus* (reviewed by Mehlin and Daneholt 1993; Wieslander 1994) offers unique possibilities for direct studies of the behavior of BR pre-mRNP particles. Due to their extraordinary size, BR particles can be unambiguously identified in the EM, and their assembly, transport, and disassembly can be directly visualized (Skoglund et al. 1983; Mehlin et al. 1992). Nascent BR RNA molecules rapidly bind proteins to form growing RNP fibers, which can be observed along the active BR genes. In the proximal parts of the BR genes, the nascent BR particles are observed as RNP fibers of increasing length. As transcription progresses, the distal (5') end of the fibers becomes packed into a dense globular structure of increasing diameter (Skoglund et al. 1983). After-transcription termination, mature BR particles are released from the chromosome and can be observed in the nuclear sap as granules with a diameter of about 50 nm (Skoglund et al. 1983). The BR particles are transported to the nuclear envelope where they become rod-shaped upon translocation through the nuclear pores (Mehlin et al. 1991).

• Electron Tomography Applications to the Balbiani Ring Transcription Products
We have used the ET technique to reconstruct in 3-D several BR RNP particles to 8.5 nm resolution from thin sections of plastic-embedded glands (Skoglund et al. 1986). An average 3-D structure was calculated from individual 3-D reconstructions that correlated with each other at an average correlation coefficient level of 0.8 (Skoglund et al. 1986). The BR RNP granules were shown to be roughly spherical with a diameter of 50 nm in the nuclear sap. They are built from a 30–60 nm wide and 10–15 nm thick RNP ribbon of the folded constituent 7 nm RNP fiber. The wide ribbon curves to form a skewed torus, or doughnut-shaped ring, so that the start and end points of the ribbon meet. Four characteristic domains can be assigned to structurally characteristic features that can be seen and identified also in the growing particle. Thus, growth of the particle is continuous and consecutive from one end to the other. Accordingly, domain 1 contains the 5' end and the 3' end is in domain 4. Consequently, the 3' and 5' ends are in close proximity to each other in the folded particle.

The translocation of the BR RNP particle through the nuclear pore has also been analysed with the ET method (Mehlin et al. 1992). For RNP particles in an early stage of nuclear pore translocation, all our ET reconstructions show that domain 1 (containing the 5' end) is fed into the nuclear pore complex first. The translocation process seems to start with a swelling of one specific part of domain 1. The rest of domain 1 subsequently changes its structure by "stretching" and enters the pore. The particle is translocated in a consecutive order, making domain 2 follow domain 1. The fact that the 5' end is transported to the cytoplasm first is also logical from a biological point of view, since the 5' end first binds to the ribosomes. The results indicate that the 5' end with its cap structure could be important both for the recognition of the particle at the nuclear pore complex and for the translocation of the particle through the pore.

We have noted that when the particle is positioned in front of the pore, and subsequently passes with the 5' end in the lead, its 3' end seems to rotate at the rim of the pore (Mehlin et al. 1995). It is still unclear to what extent the 3' poly-A region is important during the translocation per se; our data so far indicate that the 3' region has no binding specificity at this stage.

Comments

The ET technique for 3-D reconstruction gives a density which has not been iteratively improved. Iterative refinements using gradient methods or POCS (projection onto convex sets) techniques would probably only marginally improve this initial 3-D density reconstruction. A substantial improvement, however, can be achieved by running the COMET (constrained maximum entropy tomography) procedure as a post-refinement step (Skoglund et al. 1996).

The quality and reproducibility of the ET 3-D reconstruction must be assessed from at least two points of view. Firstly, one has to consider the optical resolution. In general, this is given by a formula (see the above discussion of lowpass filtering). Missing data at high tilts give a somewhat deteriorated resolution in the beam direction. Using isolated material, with a low background of, e.g., cell debris, attempts have been made to reach 2.5 nm resolution. In situ 3-D reconstructions are more complex, and so far resolution claims have stayed in the range of 4–6 nm. Secondly, the resolution at the specimen level has to be estimated. This can vary quite considerably and special care must always be taken to avoid artefacts during specimen preparation. Finally, it should be recalled that the resolution limit of stained specimens is given by the accuracy of the stain. For plastic-embedded specimens that have been positively stained, a probable resolution limit is around 2 nm.

ET 3-D reconstructions are in principle calculated as single particle reconstructions and thus they have a high noise/signal ratio. This can be improved by averaging (see above discussion of the correlation procedures). With isolated material one can probably reach the stain limit after extensive averaging, but with in situ material other factors will often degrade the final resolution. For example, a complicated distribution of fibrous material can impair the accuracy of the alignment of similar structures with the correlation methods. This could be overcome, partly, by using many more tilts than necessary from the resolution point of view, but at the cost of more specimen radiation damage.

References

Chung SY, Wooley J (1986) Set of novel, conserved proteins fold pre-messenger RNA into ribonucleosomes. Proteins: Structure, Function and Genetics 1:195–210

Daneholt B (1992) The transcribed template and the transcription loop in Balbiani rings. Cell Biol Int Rep 16:709–715

Darnell JE Jr. (1982) Variety in the level of gene control in eukaryotic cells. Nature 297:365–371

Dreyfuss G, Matunis MJ, Piñol-Roma S, Burd C (1993) HnRNP proteins and the biogenesis of mRNA. Annu Rev Biochem 62:289–321

Dworetzky SI, Feldherr CM (1988) Translocation of RNA-coated gold particles through the nuclear pores of oocytes. J Cell Biol 106:575–584

Mehlin H, Daneholt B, Skoglund U (1992) Translocation of a specific pre-messenger ribonucleoprotein particle through the nuclear pore studied with electron microscope tomography. Cell 69:605–613

Mehlin H, Skoglund U, Daneholt B (1991) Transport of Balbiani ring granules through nuclear pores in *Chironomus tentans*. Exp Cell Res 193:72–77

Mehlin H, Daneholt B (1993) The Balbiani ring particle: a model for the assembly and export of RNPs from the nucleus? Trends Cell Biol 3:443–447

Mehlin H, Daneholt B, Skoglund U (1995) Structural interaction between the nuclear pore complex and a specific translocating RNP particle. J Cell Biol 129:1205–1216

Miller OL Jr, Bakken AH (1972) Morphological studies of transcription. Acta Endocrin 168 (Suppl):155–177

Skoglund U, Andersson K, Björkroth B, Lamb MM, Daneholt B (1983) Visualization of the formation and transport of a specific hnRNP particle. Cell 34: 847–855

Skoglund U, Andersson K, Strandberg B, Daneholt B (1986) Three-dimensional structure of a specific pre-messenger RNP particle established by electron microscope tomography. Nature 319:560–564

Skoglund U, Öfverstedt L-G, Burnett RM, Bricogne G (1996) Maximum-entropy three-dimensional reconstruction with deconvolution of the contrast transfer function: A test application with adenovirus. J Struct Biol 117:173–188

Stevens BJ, Swift H (1966) RNA transport from nucleus to cytoplasm in *Chironomus* salivary glands. J Cell Biol 31:55–77

Wieslander L (1994) The Balbiani ring multigene family: coding repetitive sequences and evolution of a tissue specific cell function. Prog Nucleic Acid Res Mol Biol 48:275–313

Chapter 5

Purification and Electron Microscopy of Spliceosomal snRNPs

Berthold Kastner*

Introduction

The main focus of this chapter is electron microscopy (EM) of RNP complexes using the negative staining technique. This technique is fast and relatively easy to perform and is adequate for imaging with a standard transmission electron microscope. The prerequisite for EM analysis is the availability of isolated intact RNP particles with a concentration not less than 10–20 µg/ml. The lower size limit of the particles that can be studied by the method depends very much on their actual shape. This chapter deals with the spliceosomal RNP subunits, called small nuclear ribonucleoproteins (snRNPs). Because of their sizes, which range from about 300 kDa (U1snRNP) to up to 1500 kDa (U4/U6.U5 tri-snRNP), the snRNPs are ideal objects for EM analysis. However, snRNP protein subcomplexes as small as ~60 kDa have also been investigated successfully by negative staining EM. EM analysis is greatly facilitated if the particles studied are available as highly purified samples. Therefore, isolation procedures for snRNPs are described here in some detail.

The spliceosomal snRNPs, like many other complex macromolecular particles, are sensitive entities, and harsh purification conditions can lead to disintegration or fragmentation. Therefore, gentle methods are employed in order to isolate the snRNPs with the most complete set of specifically bound proteins. Dissociated snRNPs, on the other hand, can be valuable tools for snRNP structural analysis when defined subcomplexes are formed. For example, comparison of EM images of complete (or less disintegrated) particles with images of disintegrated parti-

* Institute of Molecular Biology and Tumor Research, Philipps Universität Marburg, Emil-Mannkopff-Str. 2, 35037 Marburg, Germany;
Tel.: (+49)-6421–28–5064; Fax: (+49)-6421–28–7008;
e-mail: Kastner@imt.uni-marburg.de

cles might lead to the identification of the structure (or structures) of the compounds only present in more complete particles. Also, in vitro reconstitution of snRNP complexes can be used for production of defined subcomplexes. Purification and reconstitution of snRNP subcomplexes are described later in this chapter.

Another strategy for the localization of a particular compound is site-specific labeling of the particles. Markers specific for a protein or RNA site can be used as long as the marker is visible by EM when bound to the particle. In this chapter procedures for labeling protein or RNA with IgG antibodies, as well as for labeling RNA sequences via complementary biotinylated oligonucleotides, are described. The speed of the negative staining procedure offers an advantage for these types of experiments, however, since finding the proper labeling conditions can be time-consuming.

Introduction to Spliceosomal snRNPs

The spliceosome is the catalytic entity that removes the introns from the primary transcripts in eukaryotes. The spliceosome consists of four small nuclear ribonucleoproteins, called U1, U2 U4/U6 and U5 snRNP, and numerous non-snRNP proteins. Each snRNP consists of one (U1, U2 and U5 snRNP) or two (U4/U6 snRNP) RNA molecules and a large number of different proteins. Within the U4/U6 snRNP, the two RNA molecules are bound to each other by extensive base-pairing. For catalysis of the splicing reaction, the snRNPs assemble together with non-snRNP proteins in an ordered manner onto the intron that is to be excised, thus forming the functional spliceosome. The spliceosome assembly pathway is shown schematically in Fig. 5.1. U1 snRNP assembles first with the 5' splice site of the pre-mRNA, followed by the association of U2 snRNP with the branch point region. Before binding to the U1-U2-pre-mRNA complex, the U4/U6 and U5 snRNPs associate with each other to form the [U4/U6.U5] tri-snRNP. The U4/U6 RNA interaction is disrupted within the spliceosome, leaving U6 RNA sequences available to

Fig. 5.1. Pathway of spliceosome assembly. The snRNPs are drawn in the ▶ shape they exhibit in the EM. In the spliceosome and the pre-spliceosomal complexes, the relative orientation of the snRNPs were arbitrarily chosen

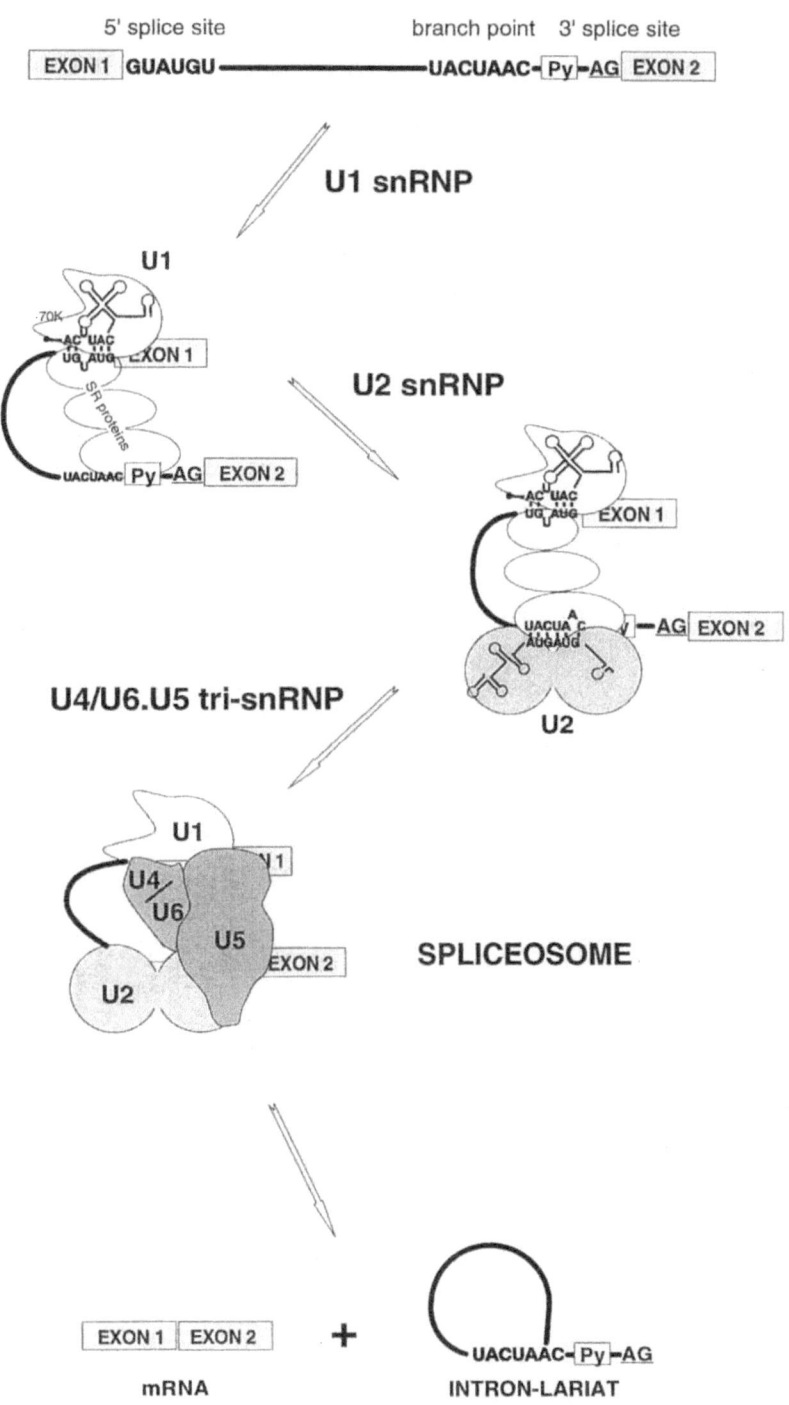

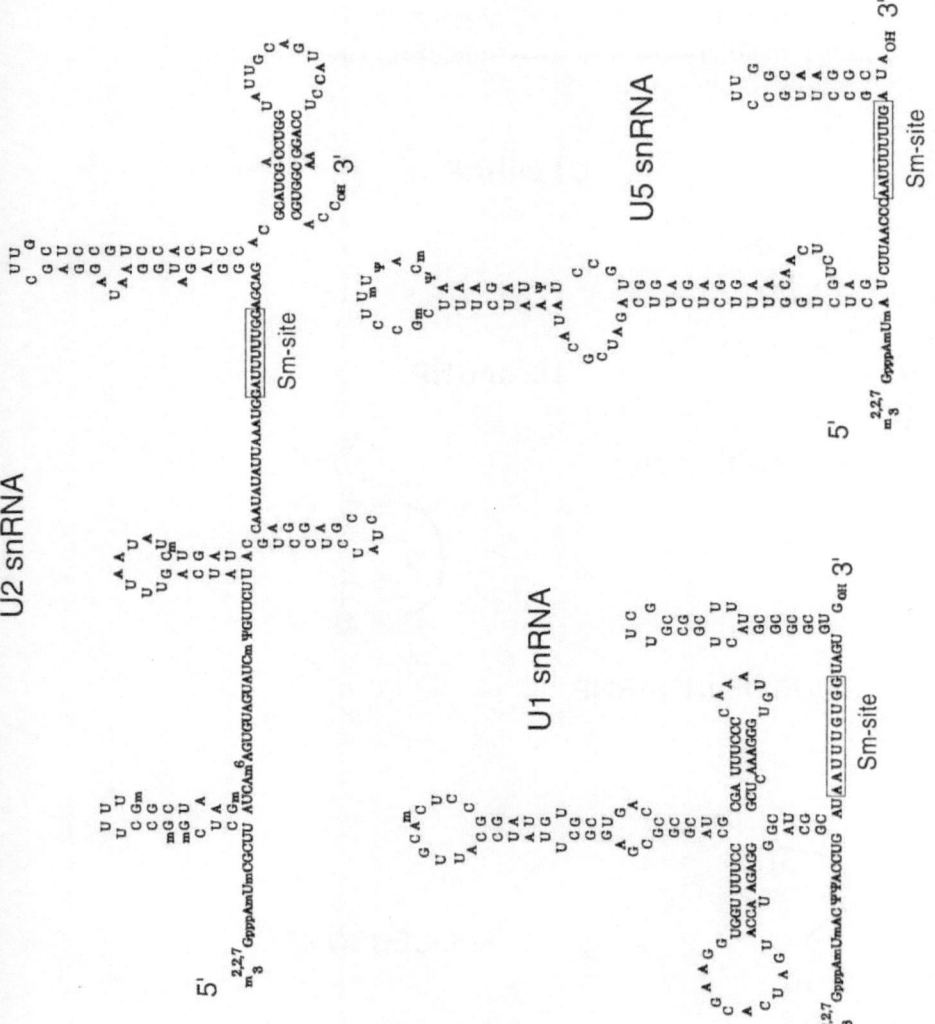

Fig. 5.2. Sequences of the human snRNAs. The sequences are drawn according to the assumed RNA secondary structures. (Guthrie and Patterson 1988)

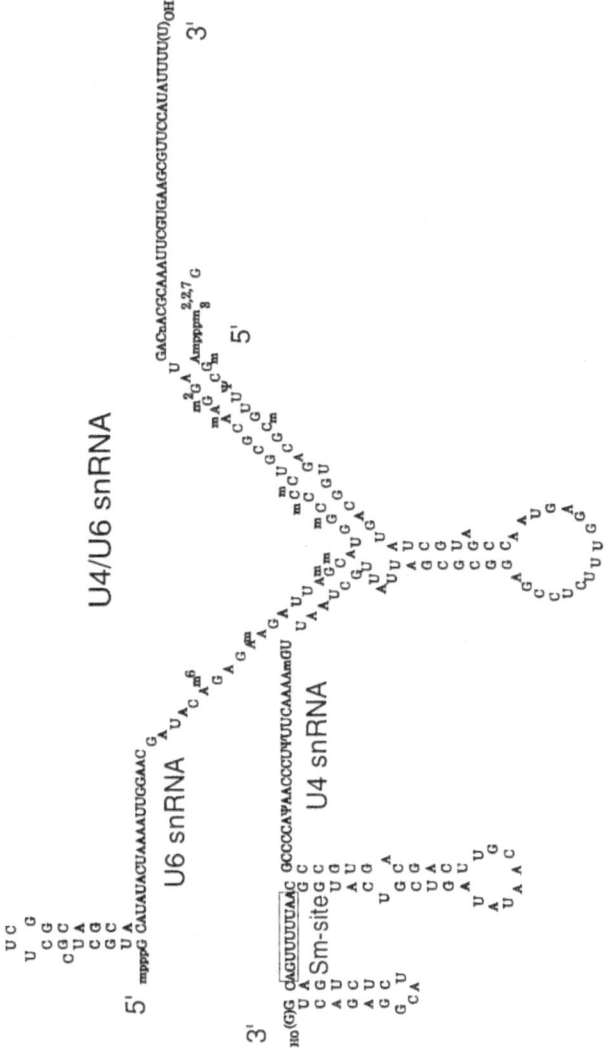

Fig. 5.2 continued

base-pair with U2 RNA as well as with intron sequences near the 5' splice site. The U5 snRNA interacts with both intron-flanking exons close to the 5' and 3' splice sites, thereby facilitating alignment of the two exons for ligation. Both of these interactions with the U5 snRNA involve participation of the loop I sequence. The interactions of U5 and U6 RNA with the 5' splice site replace the U1 RNA interaction prior to the first splice reaction. A network of RNA-RNA interactions is formed in the spliceosome, bringing the 5' and 3' splice sites, the branch point and the pre-mRNA binding sequences of U2, U5 and U6 RNA (and possibly also U1 RNA; Ast and Weiner 1996) into close spatial proximity (for reviews see Moore et al. 1993; Madhani and Guthrie 1994; Newman 1994; Nilsen 1994; Will et al. 1995).

As illustrated above, snRNA fulfils important functions in splicing. The mammalian snRNAs are relatively short with lengths between 106 (U6) and 187 (U2) nucleotides. In Fig. 5.2 the sequences of the human snRNAs are shown in their possible secondary structures (Guthrie and Patterson 1988). With the exception of U6 snRNA all spliceosomal snRNAs have a conserved motif, called the Sm site. Another hallmark of U1, U2, U4 and U5 snRNA is the characteristic trimethylguanosine-cap (m_3G-cap) at their 5' ends.

The snRNAs alone are not sufficient for splicing but must be complexed by proteins as snRNP particles. The protein compounds of the snRNPs can be divided into two classes: (1) the common proteins (B/B', D1, D2, D3, E, F, G), which are constituents of all the snRNPs, and (2) the specific proteins, which are specifically bound to only one kind of snRNP (for review, see Lührmann et al. 1990). The protein composition of the mammalian snRNPs is shown schematically in Fig. 4.3. The U1snRNP with the three specific proteins 70K, A and C is the smallest spliceosomal RNP subunit. The U2 snRNP isolated under low salt conditions contains 12 specific proteins. According to its sedimentation coefficient it is called 17S U2 snRNP (Behrens et al. 1993). At high salt concentrations (>200 mM KCl) most of the specific proteins dissociate from the U2 snRNP particle, leaving only the A' and B" U2-specific proteins. This U2 particle sediments with 12S. The U5 snRNP, with its sedimentation coefficient of 20S, is the largest single snRNP particle, containing nine (some very large) specific proteins (Bach et al. 1989). The 20S U5 snRNP is stable up to about 500 mM KCl. Above this salt concentration, it loses all its specific proteins so that only the common proteins remain. Such a complex of the common proteins with snRNA is called a snRNP core particle, and thus the common proteins are also called the core proteins. The Sm site of the snRNA is the binding site for the core proteins. Experimentally, core particles can be generated from every snRNP containing the common proteins in the presence of a strong ion-exchanger at high salt concentration and elevated temperatures (Bach et al. 1990b).

	NAME	app.M$_R$ kDa	Presence in snRNP particles				
			12S U1	17S U2	20S U5	12S U4/U6	25S U4/U6.U5
common core proteins	G	9	○	○	○	○	○
	F	11	○	○	○	○	○
	E	12	○	○	○	○	○
	D1	16	○	○	○	○	○
	D2	16,5	○	○	○	○	○
	D3	18	○	○	○	○	○
	B	28	○	○	○	○	○
	B′	29	○	○	○	○	○
U1 snRNP-specific proteins	C	22	●				
	A	34	●				
	70K	70	●				
U2 snRNP-specific proteins	B″	28,5		○			
	A′	31		○			
	SF 3A {	110		●			
		60		●			
		66		●			
	SF 3B {	53		●			
		120		●			
		150		●			
		160		●			
		33		●			
		35		●			
		92		●			
U5 snRNP-specific proteins		15			●		●
		40			●		●
		52			●		●
		100			●		●
		102			●		●
		110			●		●
		116			●		●
		200			●		●
		220			●		●
[U4/U6.U5] snRNP-specific proteins		60					●
		90					●
		15,5					●
		20					●
		27					●
		61					●
		63					●

Fig. 5.3. Protein composition of HeLa snRNPs. Each *dot* indicates the presence of the protein listed on the *left* within the snRNP at the *top* in that *column* (Lührmann et al. 1990; Behrens and Lührmann 1991; Behrens et al. 1993). Several U2 snRNP-specific proteins form RNA-free complexes; these are the A′-B″ (Scherly et al. 1990), the SF3a, and the SF3b protein complexes. (Brosi et al. 1993a)

The Spliceosomal snRNP subunits

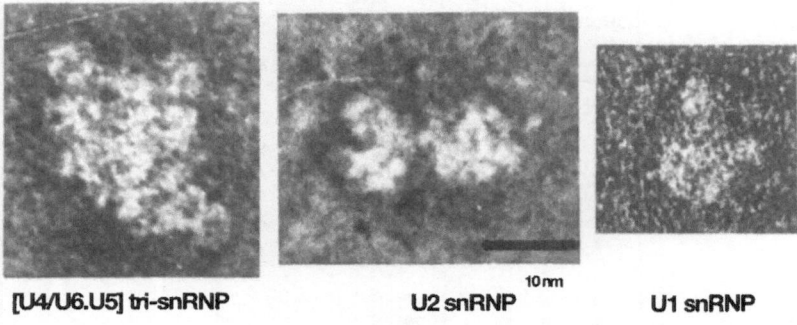

[U4/U6.U5] tri-snRNP   **U2 snRNP**   **U1 snRNP**

10 nm

The RNP subunits of the
[U4/U6.U5] tri-snRNP

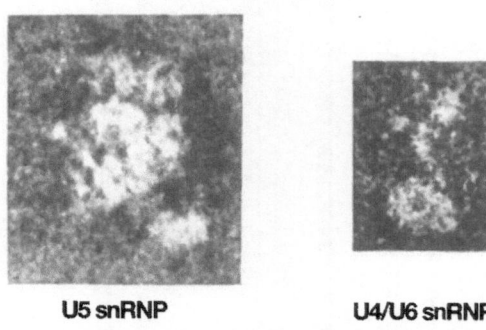

a   **U5 snRNP**   **U4/U6 snRNP**

Fig. 5.4a, b. Structures of the human snRNPs as determined by electron microscopy (EM). **a** Representative electron micrographs of negatively stained U1, 17S U2, U4/U6 and 20S U5 snRNPs as well as the [U4/U6.U5] tri-snRNP complex. **b** Models of the snRNPs showing the positions of components localized by EM (see pages 103 and 104)

At low salt concentrations the U5 and the U4/U6 snRNPs associate with a set of five additional proteins to form the 25S [U4/U6.U5] tri-snRNP, which is the functional unit of these two snRNPs (Behrens and Lührmann 1991). Exposure to more than 400 mM salt leads to dissociation of the tri-snRNP into a 12S U4/U6 snRNP and the 20S U5 snRNP.

All four spliceosomal snRNPs as well as the [U4/U6.U5] tri-snRNP complex have been isolated from HeLa cells, and their structures have been studied by EM using negatively stained specimens (Kastner and Lührmann 1989; Kastner et al. 1990,

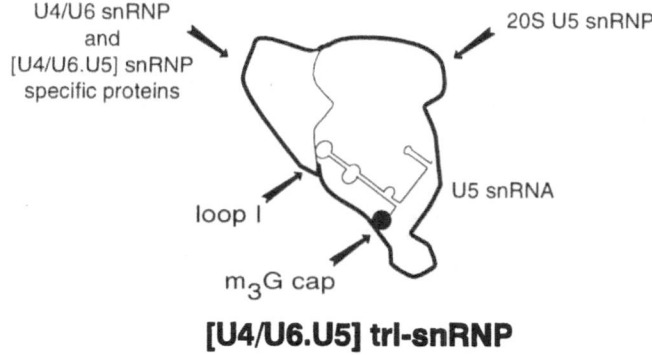

[U4/U6.U5] tri-snRNP

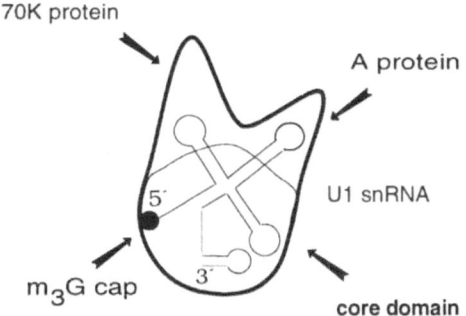

U1 snRNP

Fig. 5.4b.

1991; Behrens et al. 1993). As shown in Fig. 5.4a, each particle has a characteristic asymmetric structure with dimensions ranging from 8 nm (width of the body of the U1 snRNP) to about 25 nm (length of the [U4/U6.U5] tri-snRNP particle). The U1 snRNP has a structure consisting of a roughly round main body and two small, adjacent protuberances. A main body, similar to that of U1 snRNP, can also be seen at the U4/U6 snRNP. Here, in addition, there is a filamentous Y-shaped domain protruding from the body. The 17S U2 snRNP has two similar sized globular domains, so that the particle appears dumbbell shaped. The 20S U5 snRNP has an elongated structure with a large head, a central

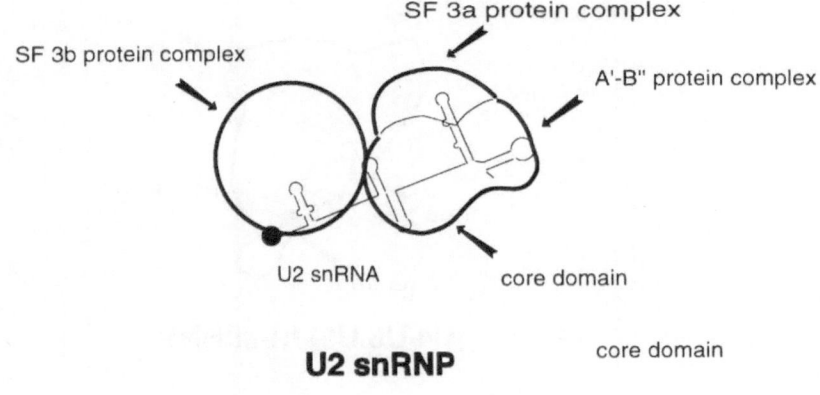

U2 snRNP

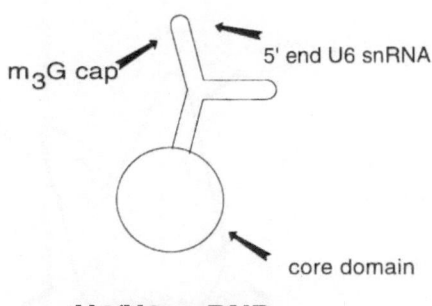

U4/U6 snRNP

Fig. 5.4b. (continued)

body, and appear in the images either with a pointed straight or a bent lower end. The [U4/U6.U5] tri-snRNP has a more tri-angle-like structure, in which the lower part shows similarities to the corresponding part of the U5 snRNP, while the upper part is much broader in the tri-snRNP particle.

By site-specific labeling or specific depletion of components, structural domains, individual proteins or RNA, sequences could be located within the particles (Kastner and Lührmann 1989; Kastner et al. 1990, 1991, 1992; Hoet et al. 1993; Gröning et al. 1997). A summary of the results on the architecture of the spliceosomal snRNPs is shown in Fig. 5.4b. The locations of the first and second stem/loop of U1 RNA as well as the last stem/loop of U2 RNA are deduced from localization of proteins which bind to these RNA structures, which are the A and 70K, and A' and B" proteins, respectively. Some functionally important RNA sites, such as the 5' end of U1 RNA, which base-pairs with the 5' splice site, and the exon bridging loop I of U5 RNA, have been located by specific RNA-labeling.

5.1
Purification of snRNPs from HeLa Cells

The snRNP purification protocol described here was developed in the laboratory of Reinhard Lührmann (Lührmann et al. 1982; Bringmann et al. 1983; Bringmann and Lührmann 1986; Bochnig et al. 1987; Bach et al. 1989; Behrens and Lührmann 1991; Behrens et al. 1994). Two features of the snRNPs make their isolation difficult: (1) The low abundance of the snRNPs in the cell. In HeLa cells there are about 10^6 copies of U1 snRNP and even less of the other snRNPs. (2) Most of the protein compounds are associated with the snRNPs in a labile fashion, and exposure to high salt concentration results in dissociation of most of the specific proteins, as discussed above.

To overcome these problems, affinity chromatography with competitive elution was performed, allowing a high enrichment of all spliceosomal snRNP particles in one step under very mild conditions. In combination with density gradient centrifugation, the very labile 17S U2 snRNP and the [U4/U6.U5] tri-snRNP can be purified to high homogeneity (Behrens and Lührmann 1991; Behrens et al. 1993). For the more stable snRNP particles, such as the U1 snRNP, the 12S U2 snRNP, and the U4/U6 snRNP, ionexchange chromatography is additionally employed (Bach et al. 1990a). Figure 5.5 illustrates the various steps for obtaining the different mammalian snRNP species. Below, protocols are given for preparing nuclear extracts from HeLa cells, running the affinity and the Mono Q ion-exchange columns, and performing density gradient centrifugation.

All procedures described here should be performed at 4 °C unless otherwise stated. All the samples obtained by the protocols should be used immediately for further fractionation or EM specimen preparation. If this is not possible, they should be aliquoted, frozen quickly in liquid nitrogen and stored at −80 °C.

The snRNPs isolated according to these protocols show functional activity as they can restore splice activity of nuclear extract depleted of the particular snRNP. Also given is a protocol for preparing nondenatured, RNA-free mixtures of the common

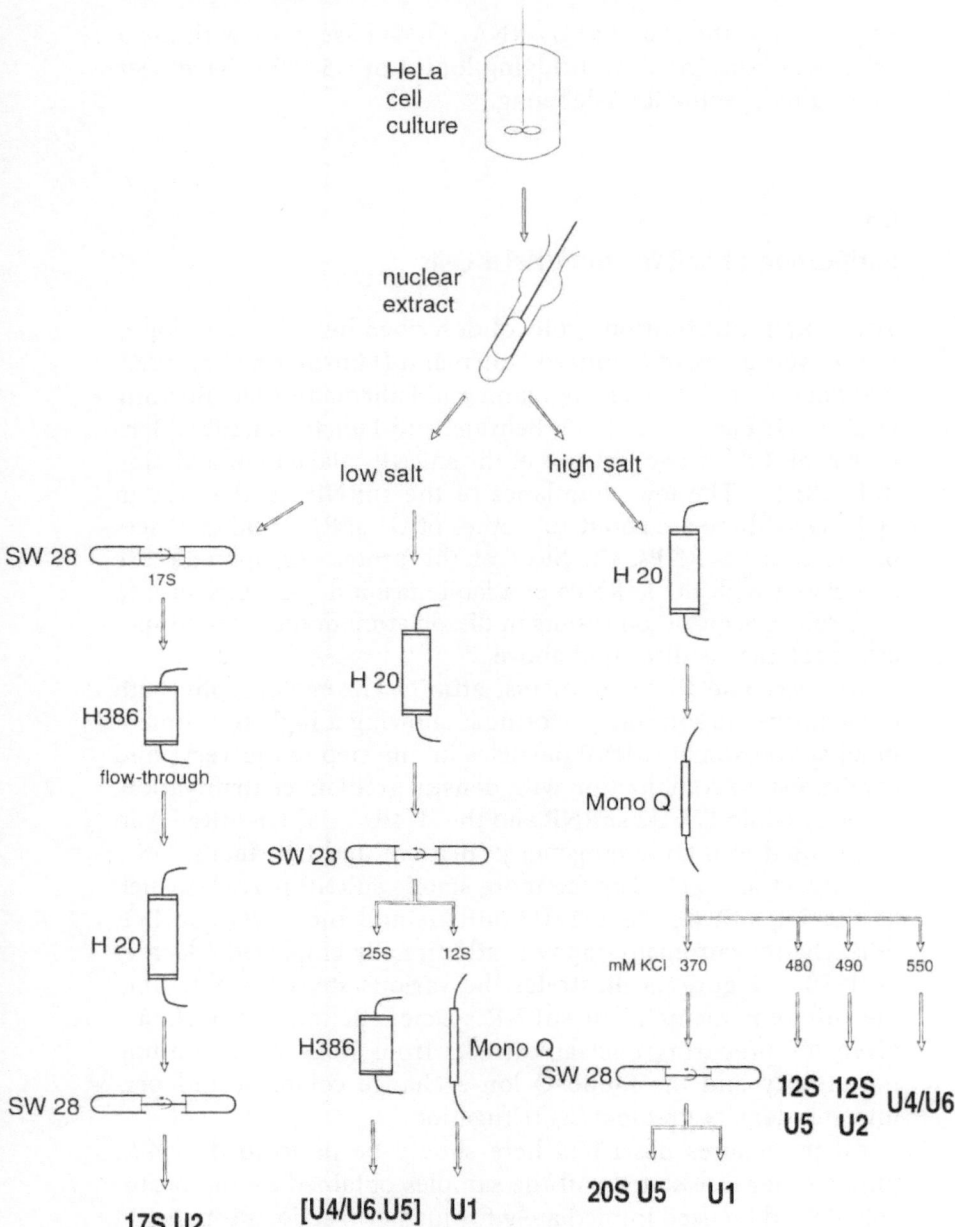

Fig. 5.5. Purification schema for mammalian snRNPs. The procedures of the various steps are shown as *symbols*. SW28 indicates preparative scale glycerol or sucrose gradient centrifugation, Mono Q stands for Mono Q chromatography, and H20 and H386 indicate anti-m₃G and anti-70K/ 100 kDa immunoaffinity chromatography, respectively

snRNP proteins. These proteins can be used for further fractionation by sucrose gradient centrifugation (Raker at al. 1996) or for in vitro reconstitution of functionally active snRNP cores (Sumpter et al. 1992; Ségault et al. 1995). Some procedures for these methods have been described in detail previously (Bach et al. 1990a; Will et al. 1993; Behrens et al. 1994).

For HeLa cells additional purification schemes for obtaining nondenatured RNP complexes have been developed in other laboratories. The purification scheme developed in Angela Krämer's laboratory focuses in particular on the isolation of protein splicing factors (Krämer 1990 and 1992; Krämer and Utans 1991; Brosi et al. 1993b), while the protocols developed in Robin Reed's laboratory aim at the isolation of spliceosomes and prespliceosomal complexes (Reed et al. 1988; Bennett et al. 1992; see also Furman and Glitz 1995). In the laboratories of Ruth Sperling and Josef Sperling, methods have been developed for the isolation of spliceosomal subunits containing large nuclear RNPs (lnRNPs) (Spann et al. 1989; Miriami et al. 1995).

Preparation of Nuclear Extracts from HeLa Cells

The snRNPs are most concentrated in the cell nucleus, the compartment in which they fulfill their task of pre-mRNA splicing. Therefore snRNPs are preferably isolated from the nucleus. For HeLa cells, isolation of nuclei and preparation of splicing-active nuclear extracts is a standard procedure in many laboratories. Originally, the protocol was developed for preparing extracts active in polymerase II transcription in vitro (Dignam et al. 1983). SnRNPs can also be isolated in principle from total cell extracts, as shown recently for the yeast *Saccharomyces cerevisiae* (Fabrizio et al. 1994).

Procedure

1. HeLa cells can be obtained by growing the cells in bottles or in a bioreactor, or they can be purchased commercially. Frozen cells or prepared nuclear extracts suitable for snRNP isolation can be obtained from Computer Cell Culture Centre (4C, Mons, Belgium). To grow HeLa S3 cells, the cells must be kept at a density between 2.5 and 5×10^5/ml medium at logarithmic growth rate in suspension culture in S-MEM

HeLa nuclear extract

(Gibco, BRL Life Technologies) supplemented with 5 % (v/v) newborn calf serum (Gibco, BRL Life Technologies), 50 µg/ml penicillin (Boehringer Ingelheim Bioproducts), and 100 µg/ml streptomycin (Boehringer Ingelheim Bioproducts) at 37 °C. At least 5×10^9 cells should be accumulated for a harvest.

2. Harvest the cells by centrifugation in a Heraeus Cryofuge 6000 with swinging bucket rotor for 10 min at $1000\,g$. Alternatively, smaller rotors such as the Sorvall HB4 can be used with several successive centrifugations.

3. Resuspend the cells with 20 ml PBS-Earl (130 mM NaCl; 20 mM K_2HPO_4/KH_2PO_4, pH 7.4) per 10^9 cells and pellet in a Sorvall HB4 rotor for 10 min at $1000\,g$.

4. Determine the volume of the cell pellet and resuspend in five volumes of buffer A (10 mM Hepes-KOH, pH 8; 10 mM KCl; 1.5 mM $MgCl_2$; 0.5 mM dithioerythritol, DTE).

5. Let the cells swell for 10 min, pellet again, and resuspend in two volumes of buffer A.

6. Lyse the cells by 10 strokes of the 40 ml Dounce homogenizer (Kontes Glass).

7. Separate the nuclei from the cytoplasm by two successive 10 min centrifugations in a Sorvall SS 34 rotor, first at $1000\,g$ and then at $25\,000\,g$.

8. Resuspend the nuclei in 3 ml buffer C (20 mM Hepes-KOH, pH 8; 420 mM NaCl; 1.5 mM $MgCl_2$; 0.5 mM DTE; 0.5 mM PMSF; 0.5 mM EDTA, pH 8, 25 % (v/v) glycerol) per 10^9 cells.

9. Open the nuclei by 10 strokes of the 40 ml Dounce homogenizer.

10. Transfer the suspension into a beaker and stir carefully with a stir-bar on ice for 30 min.

11. Remove the nuclear membrane by centrifugation in a SS 34 rotor for 30 min at $25\,000\,g$.

12. Collect the supernatant, which is the nuclear extract.

The salt concentration of the nuclear extract is now about 250 mM. For isolation of the salt-sensitive 17S U2 snRNP and [U4/U6.U5] tri-snRNP, the salt concentration of the nuclear extract and the buffers used for the further purification steps should be kept at 250 mM or lower (low salt procedure), while for isolation of the stable U1 snRNP and partially disintegrated snRNPs, salt concentrations up to 450 mM should be used (high salt procedure). Nuclear extract active in splicing in vitro can be obtained by dialysis with buffer G (20 mM Hepes-KOH, pH 8; 150 mM KCl; 1.5 mM $MgCl_2$; 0.5 mM DTE; 0.5 mM PMSF; 5 % (v/v) glycerol).

Immunoaffinity Chromatography

The snRNP purification scheme (Fig. 5.5) employs two immunoaffinity chromatography columns: the H20 column, with a bound m_3G-cap specific monoclonal IgG antibody (H20) and the H386 column, with a bound monoclonal IgM antibody reactive with the U1 snRNP-specific 70K protein as well as with the U5 snRNP-specific 100 kDa protein (H386). Once adsorbed, snRNPs can be eluted by competing with free m^7G nucleotide (in the case of the H20 column) or with cross-reactive synthetic peptides (in the case of the H386 column).

In principle, other monospecific antibodies could be used if: (1) they are highly specific for snRNPs, (2) the epitope is known and can be synthesized, and (3) the kinetics of the interaction allow efficient retention as well as competitive elution.

Depending on the salt concentration used the H20 anti-m_3G affinity column can be run in two ways: (1) with buffer C_a (20 mM Hepes-KOH, pH 8; 420 mM KCl; 1.5 mM $MgCl_2$; 0.5 mM DTE; 0.5 mM PMSF; 0.5 mM EDTA, pH 8; 5 % (v/v) glycerol) for isolation of the stabile and partially disintegrated snRNPs with the high salt procedure, and (2) with buffer C_b (same as C_a, but with 250 mM KCl) for isolation of the salt-labile snRNPs with the low salt procedure.

H20 column

1. Equilibrate a 5 ml anti-m_3G immunoaffinity column by washing with approximately five column volumes of buffer C_a (or C_b, as required).

2. Clear nuclear extract by centrifugation in a Beckman Ti70 rotor at 165 000 g for 30 min and subsequent filtration of the supernatant through a 5 μm membrane filter, followed by a 1.2 μm filter. Dilute nuclear extract prepared from 5×10^9 HeLa cells to ~25 ml with buffer C_a (C_b).

3. Apply the diluted nuclear extract prepared from 5×10^9 HeLa cells to the affinity column at about 1.5 ml/h. For isolation of 17S U2 snRNPs, apply the flow-through of the H386 anti-70K/100 kDa column.

4. Elute nonspecifically bound components of the extract with about six column volumes of buffer C_a (C_b). For isolation of 17S U2 snRNPs, the buffers in this and the following step should contain only 150 mM KCl.

5. Elute the specifically bound snRNPs using 15 mM m^7G nucleoside dissolved in buffer C_a (C_b). Collect 1 ml fractions and determine the protein concentrations of the fractions by the method of Bearden (1978). When unfractionated nuclear extract from 5×10^9 HeLa cells is loaded onto the column, 2–4 mg snRNPs can be eluted. Analyze the protein and RNA compositions of the fractions by polyacrylamide gel electrophoresis (PAGE) (Will et al. 1993).

6. Remove the antibody-bound m^7G nucleoside by washing the column with buffer C_a supplemented with 6 M urea.

7. Regenerate the affinity column by washing with 20 column volumes of C_a. For long-term storage, add NaN_3 to give a final concentration of 0.02 %.

H386 column

1. Equilibrate a 2 ml H386 anti-70K/100 kDa immunoaffinity column by washing with about five column volumes of buffer G (20 mM Hepes-KOH, pH 8; 150 mM KCl; 1.5 mM $MgCl_2$; 0.5 mM DTE; 0.5 mM PMSF; 5 % (v/v) glycerol).

2. Pool the 17S gradient fractions for isolation of 17S U2 snRNP, and the 25S fractions for isolation of 25S [U4/U6.U5] tri-snRNPs, and apply to the H386 column at 1 ml/min. About 100–150 µg snRNPs can be loaded onto a 2 ml column.

3. For isolation of 17S U2 snRNPs, collect the flow-through and load onto an H_2O (anti-m_3G) column.

4. Elute the nonspecifically bound components from the H386 column with about 20 column volumes of buffer G.

5. For isolation of [U4/U6.U5] tri-snRNPs elute the specifically bound snRNPs with five column volumes of a 0.01 mM solution of a competing peptide in buffer G (the primary epitope of the H386 antibody is contained in the 32-mer peptide, DRDRERRRSHRSERERRRDRDRDRDRDREHKR; see Beh-

rens and Lührmann 1991). Collect 500 µl fractions and analyze one-tenth of each fraction for RNA and protein content
by PAGE followed by Coomassie or silver staining (Will et al.
1993). For further purification of the [U4/U6.U5] tri-snRNP
and removal of the excess peptide, load the appropriate fractions onto a glycerol or sucrose gradient.

6. Elute the antibody-bound peptide with five column volumes
   of phosphate buffer (10 mM K_2HPO_4/KH_2PO_4, pH 7.2) and
   then with five column volumes of 3.5 M $MgCl_2$ in the phosphate buffer.

7. Regenerate the affinity column by washing with ten column
   volumes of buffer G. For long-term storage, add NaN_3 to give
   a final concentration of 0.02 %.

Density Gradient Centrifugation

Due to the large size differences of the spliceosomal snRNPs,
fractions enriched in particular snRNP species can be obtained
by either glycerol or sucrose gradient centrifugation. Usually
fractions containing the 25S, 20S, 17S or the 12S snRNPs can be
obtained by gradient centrifugation. Glycerol gradients are his-
torically used for separation of splicing complexes (Frendeway
and Keller 1985; Grabowski et al. 1985). However, sucrose gradients are recommended for EM analysis, as glycerol has a relatively low evaporation temperature. Traces of glycerol trapped
between the carbon films of EM specimens (see below) evapo-
rate easily in the high vacuum of the EM and can sometimes produce artefacts. Sucrose, on the other hand, has a positive effect
on EM sample preparation by promoting an even background
staining.
     Depending on the snRNPs to be separated, different salt con-
centrations should be used for the gradients. To obtain the high
molecular weight, labile snRNPs, it is necessary to run the gradient under low salt conditions to prevent protein dissociation.
The stabile snRNPs, by contrast, can be centrifuged at higher salt
concentrations. Centrifugation at a low salt concentration has
the disadvantage that weak interactions between the snRNPs are
promoted, often resulting in a broadening of the peaks. In particular U1 snRNP "smears" over the entire gradient at low salt
concentrations. It is thus advantagous to use gradient buffers
with optimized salt concentrations. To separate the 20S U5

snRNPs from U1 snRNPs, gradients with a salt concentration of 250 mM should be used. To sediment the fragile 17S U2 snRNP, the salt concentration in the buffer should not be above 150 mM, while for the [U4/U6.U5] tri-snRNP, up to 200 mM can be used (see Table 5.1).

The necessity of using low salt concentrations limits the resolution of the gradient centrifugation of the 17S U2 snRNP and the [U4/U6.U5] tri-snRNP, so that a satisfactory separation is not obtained in a single centrifugation step. Consequently, the snRNP purification schema (Fig. 5.5) contains two centrifugation steps for these particles. If the snRNPs are used directly for EM analysis, the second gradient centrifugation step should be performed immediately before EM sample preparation. Dissociated proteins and smaller complexes generated by storage or freezing and thawing (as well as the m^7G nucleotide or the peptide used for competitive elutions) is separated from the larger, intact snRNPs by gradient centrifugation. Micrographs of these samples usually show a homogeneous background.

The final centrifugation step can be omitted if small contaminants do not interfere with the subsequent analysis. Immunolabeling of 20S U5 snRNP, for example, can be done also in the presence of contaminating 12S snRNPs, if a gradient centrifugation step is used subsequently for immunocomplex enrichment (see below).

Table 5.1. Parameters for density gradient centrifugation of snRNPs

Gradient solutions				Preparative scale SW 27 rotor		EM scale TLS 55 rotor	
snRNP/proteins	KCl (mM)	Sucrose		Speed (rpm)	Time (h)	Speed (rpm)	Time (h)
		Low (%)	High (%)				
12S U1	300	5	20	28 000	25	55 000	6
17S U2	150	10	30	27 000	17	55 000	4
20S U5	250	10	30	28 000	16	55 000	3
25S U4/U6.U5	200	10	30	27 000	16	55 000	2.5
Core proteins	150	5	20			55 000	12

Recommended KCl and sucrose concentrations are given for the solutions to be used for gradient formation, as well as the speed and time for preparative and EM scale centrifugations of the various snRNPs.

1. Use a Beckman SW28 or an equivalent rotor (with a Beckman L8 or an equivalent ultracentrifuge) for preparative gradient centrifugations and a Beckman TLS 55 rotor (with the Beckman table-top ultracentrifuge TLA 100) for EM sample preparation.

Sucrose gradient centrifugation

2. Prepare the low and high gradient solutions in gradient buffer (20 mM Hepes-KOH, pH 8; 1.5 mM MgCl$_2$; 0.5 mM DTE; 0.5 mM PMSF) with the desired KCl and sucrose concentrations as specified in Table 5.1. For UV monitoring of very small amounts of snRNPs during gradient fractionation (see step 5), the optical densities of the low and high gradient solutions can be matched by adding a small amount of tyrosine to the low gradient solution.

3. Pour linear gradients in the appropriate centrifuge tubes. For easy, fast and reproducible linear gradients, the BioComp Gradient Master (Fredericton, N.B., Canada) is recommended. This gradient former works especially well for the small TLS 55 gradients (1.5 ml). The functional principle of the BioComp Gradient Master is illustrated in Fig. 5.6a. Store the gradients up to one hour undisturbed at 4 °C.

4. Load the sample carefully and evenly onto the gradient with an Eppendorf pipette. If the sample contains more, or the same, concentration of glycerol or sucrose as the "low gradient solution," reduce the density of the sample by diluting with an appropriate buffer, so that the concentration is at least 2 % below that of the low density gradient solution.

5. Start centrifugation in a pre-cooled (4 °C) ultracentrifuge at a low acceleration rate, and run it as specified in Table 5.1. With the TLS55 rotor, centrifugation can be stopped without braking.

6. Harvest the gradients either manually in 15–25 equal fractions from the top using an Eppendorf pipette or automatically from the bottom with simultaneous monitoring of the optical density as described in Fig. 5.6b. Parameters for automatic fractionation are shown in Table 5.2.

7. Analyze the protein and RNA compositions of the fractions by PAGE (Will et al. 1993).

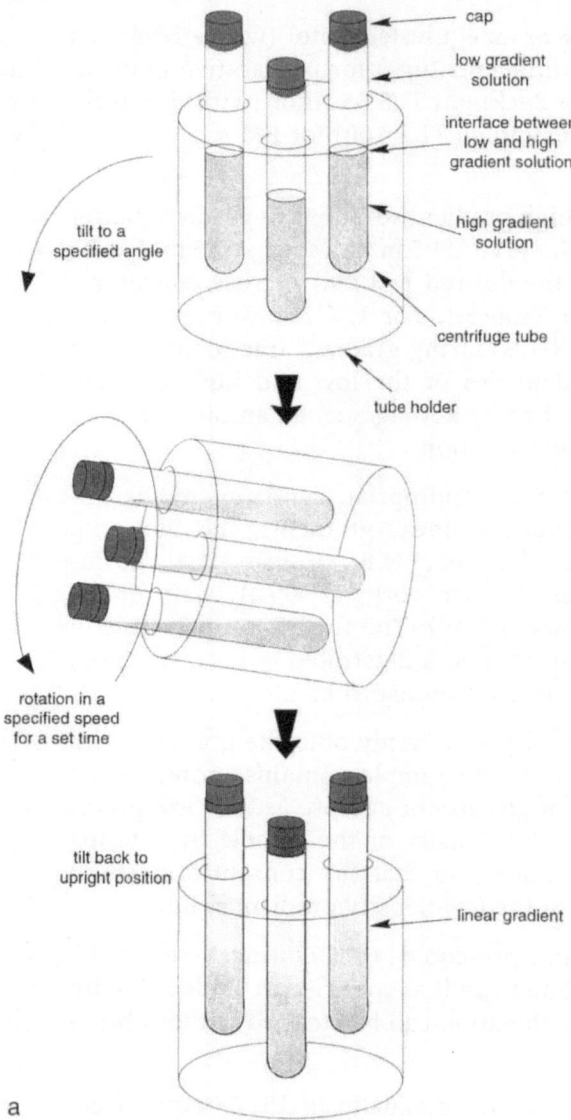

cap

low gradient
solution

interface between
low and high
gradient solution

high gradient
solution

tilt to a
specified angle

centrifuge tube

tube holder

rotation in a
specified speed
for a set time

tilt back to
upright position

linear gradient

a

Fig. 5.6a, b. Formation and fractionation of density gradients. **a** The functional principle of the BioComp gradient master is shown in three steps. First, the centrifugation tube is filled in the lower half with the high, and upper half with the low density, gradient solution, without mixing the solutions. The tubes are covered without trapping any air either with an appropriate cap, or with Parafilm, and placed into the tube holder (which holds up to six tubes). Next, the tube holder tilts and rotates. Tilt angle, rotation speed, and time are specified by the tube size and type of gradient. Finally, the tube holder returns to the upright position, finishing the gradient formation.

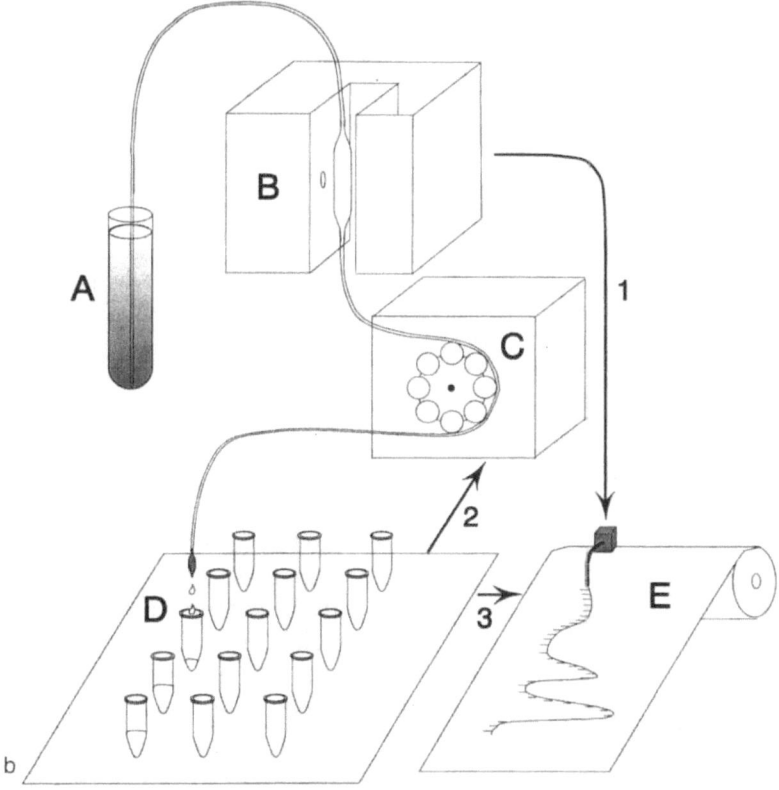

Fig. 5.6a, b. Formation and fractionation of density gradients. b The automatic gradient fractionation system. The gradient in the centrifuge tube (A) is collected from the bottom by an inserted canule. The gradient is fractionated by a peristaltic pump (C), and the optical density is measured continuously (B). A cuvette with a round cross-section is recommended. The optical density signal is transferred (1) to the recorder (E), where the signal is plotted. Fractions are collected (D) and the signal for the fraction change is monitored (3) by the recorder. The fraction change signal is also communicated to the pump (C), which, in response, pauses during fraction change. Before gradient fractionation can be started, the entire tubing system should be filled with the high gradient solution. All air trapped in the cuvette should be removed by inverting the flow direction. For this, the amount of solvent contained in the tubing between (C) and (D) must be larger then that between (A) and (C)

Table 5.2. Parameters for automatic gradient fractionation

	Preparative scale SW 27 rotor	EM Scale TLS 55 rotor
Fraction size	1.5–2 ml	3 drops
Number of fractions	23–17	~15
Pump speed	2 ml/min	0.7 ml/min
Chart speed	1 cm/min	5 cm/min

Parameters for programming the equipment are given for the fractionation of both preparative and EM scale gradients.

Ion-Exchange Chromatography

Ion-exchange chromatography is performed with Mono Q columns (Pharmacia) for the next step in snRNP purification (see scheme in Fig. 5.5). Here, the snRNP samples are fractionated with the FPLC system (Pharmacia) over a 1 ml Mono Q column; alternatively, 0.1 ml Mono Q columns can be used together with the Smart system (Pharmacia) for small amounts of sample. The Mono Q resin is a strong anion-exchanger and has an excellent resolution power. However, high salt concentrations are necessary for elution of the snRNPs, so that only snRNPs with tightly associated proteins can be purified by Mono Q chromatography. In the first peak, at about 370 mM KCl, the majority of the U1 snRNPs elute. This peak also contains most of the 20S U5 snRNP. By glycerol gradient centrifugation both snRNPs can be separated either before or after the Mono Q chromatography. Centrifugation of the U1/U5-containing fraction from Mono Q chromatography must be carried out before freezing the purified samples since the high molecular weight U5 proteins tend to dissociate, forming 12S U5 particles which will cosediment with the 12S U1 snRNPs during gradient centrifugation. A second U5 snRNP peak elutes from the Mono Q column at 480 mM KCl, whereas the majority of U2 and U4/U6 snRNPs elute at 490 mM and 550 mM KCl, respectively (Bach et al. 1990a; Will et al. 1993; Behrens et al. 1994). U1 snRNPs lacking one ore more of their specific proteins (C, A, 70K) can be isolated by performing Mono Q chromatography at higher temperatures. Small amounts of protein-deficient U1 snRNP particles are also often observed after chromatography at 4 °C (Bach et al. 1990a, b).

To reduce the degree of dissociation of the particles during chromatography, the Mono Q column can be replaced by a Resource Q column (Pharmacia). Chromatography of snRNPs with Resource Q resin allows fractionation of the labile snRNPs to a degree. Unfortunately, the resolution of the fractionation with Rescue Q is somewhat lower than that obtained with Mono Q.

1. Wash the FPLC (fast protein liquid chromatography) system, which includes a 50 ml "superloop" and a Mono Q HR 5/5 column (1 ml bed volume), with a 20-fold system volume amount of Mono Q buffer (20 mM Tris-HCl, pH 7.0; 1.5 mM $MgCl_2$; 0.5 mM DTE; 0.5 mM PMSF) containing 1 M KCl.

   Mono Q chromatography

2. Wash and equilibrate the column with 20-fold system volume amount of Mono Q buffer containing 50 mM KCl. Monitor the absorbance of the column flow-through at 280 nm. The value obtained is the zero point for subsequent absorbance measurement.

3. Dilute the snRNP sample with Mono Q buffer so that the concentration of monovalent ions is less than 200 mM.

4. Load the snRNPs (1–40 mg) onto the Mono Q column, using the superloop, with a flow rate of 2 ml/min. The pressure should not exceed 3.0 MPa.

5. Wash with Mono Q buffer containing 50 mM KCl until the fraction absorbance at 280 nm reaches zero.

6. Elute the snRNPs from the column with a flow rate of 1 ml/ min using Mono Q buffer containing 50 mM KCl (buffer A) or 1 M KCl (buffer B) and the following gradient: Start with 100 % buffer A. Increase the amount of buffer A with a velocity of 5.4 %/min for 4 min, 1 %/min for 30 min, and then 4.2 %/min for 10 min. Finish with 100 % buffer B for 4 min. Collect 1 ml fractions during the entire elution.

7. Determine the snRNP concentration in each fraction either by measuring the absorbance at 280 nm (approximately 0.35 mg/ml at $A_{280,1\ cm} = 1$ for U1 snRNP) or by the method described by Bearden (1978). Analyze the protein and RNA compositions of the fractions by PAGE (Will et al. 1993).

8. Separate the 20S U5 snRNP and the U1 snRNP from each other by glycerol or sucrose density gradient centrifugation immediately after Mono Q chromatography.

Isolation of snRNP Proteins and Reconstitution of snRNP Core Particles

In contrast to most of the specific proteins, the core proteins are stably associated with the snRNAs. Thus, dissociation of the snRNP core particle without denaturation of the core proteins requires particular conditions. For this, the disassembly method described for the signal recognition particle by Walter and Blobel (1983) has been adapted (Sumpter et al. 1992). The protein-RNA interactions within the snRNP particle are first weakened by chelation of divalent cations with EDTA. The proteins are subsequently separated from the RNA by ion-exchange chromatography over the polycationic resin DE53. Optimal recovery of snRNA-free snRNP proteins occurs when the disassembly is performed in buffer containing 150 mM K-acetate (KAc), 140 mM NaCl, and 5 mM EDTA. A monovalent cation concentration less than 300 mM is required to prevent the release of snRNA from the DE53 resin.

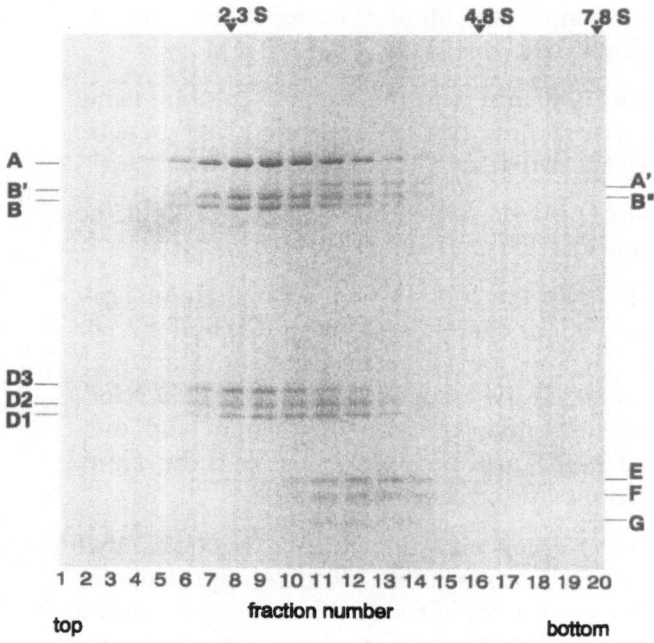

Fig. 5.7. Gradient centrifugation of snRNP proteins. The proteins were isolated from snRNPs as described and then fractionated by sucrose gradient centrifugation (5–20% sucrose, 26 h at 45 000 rpm in a TLS55 rotor). Each fraction was analyzed by PAGE. The smallest proteins, E, F and G, cosediment with the fastest sedimentation speed. (Plessel et al. 1997, reprinted by permission of the publisher Academic Press Limited London)

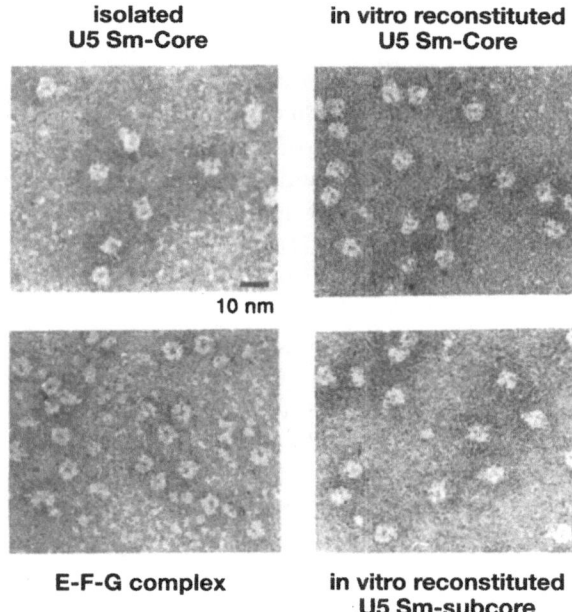

Fig. 5.8. Electron micrographs of negatively stained isolated U5 snRNP cores and in vitro reconstituted U5 snRNP cores, the E-F-G protein complex, and in vitro reconstituted U5 snRNP subcores, which lack D3-B/B' proteins

Under the conditions described in the following protocol, the snRNPs and the majority of the specific proteins bind to the DE53 resin, while all of the common proteins (B, B', D1, D2, D3, E, F, G) and the U1-specific proteins A and C, and the U2-specific proteins A' and B" remain in solution (Sumpter et al. 1992). The recovery efficiency of soluble proteins is approximately 20 %–30 % (Will et al. 1993). These proteins are present in specific protein-protein complexes, namely A'-B", D3-B/B', D1-D2, and E-F-G (Raker et al. 1996). The E-F-G complex can be separated from the other proteins by sucrose gradient centrifugation since it is the fastest sedimenting complex (with 3.7S), as shown in Fig. 5.7. The E-F-G complex is most likely a hexamer with a molecular weight of ~60 000. It has a doughnut-like appearance in the EM (Fig. 5.8) (Plessel et al. 1997).

The snRNP core proteins thus obtained can be reconstituted into intact snRNP cores by incubating with Sm site-containing snRNAs (Sumpter et al. 1992; Ségault et al. 1995). For efficient reconstitution of snRNPs, the RNA-free protein preparation must first be concentrated by dialysis against a 30 % (w/v) polyethylene glycol (PEG) buffer. This method allows a 50- to 100-fold increase in protein concentration to be consistently achieved without significant protein loss. Reconstitution is generally performed with an individual snRNA species, isolated from native snRNP particles or generated by in vitro transcription. Optimal reconstitution of snRNPs is observed when snRNA is incubated with a five- to tenfold molar excess of proteins over snRNA (for snRNP proteins, this corresponds to about 1 mg protein/pmol RNA). Reconstitution is typically performed in a buffer containing 5 mM $MgCl_2$ and 50 mM KCl, although little change in the efficiency of particle formation is observed if the former is between 2 and 15 mM and the latter between 50 and 250 mM (Will et al. 1993).

The reconstituted snRNPs can be separated from non-incorporated proteins and RNA by gradient centrifugation. EM analysis of reconstituted snRNP cores (Fig. 5.8) has demonstrated that reconstituted cores have the same structure as their counterparts directly isolated from HeLa cells (Fig. 5.8; Plessel et al. 1997). By in vitro reconstitution incomplete snRNP cores can be produced as well. A subcore lacking D3-B/B' can be obtained by immunodepletion of the D3-B'/B complex from the core protein mixture before RNP reconstitution (Raker et al. 1996). In EM, the shapes of gradient-purified U5 snRNP subcores appear very similar to those of complete cores (Fig. 5.8). However, there are differences in the structural features which indicate that the D3-B/B' domain is absent on the subcore partcles (Plessel et al. 1997).

Isolation of snRNP proteins

1. Prepare the DE53 resin (Whatman) by resuspending it in a tenfold volume of 4.0 M KAc, pH 5.5. Use 2.5 ml resin per 1 mg snRNP. Let the resin set for 5 min.

2. Wash the resin four times with a tenfold volume of sterile water and then four times with a tenfold volume of wash buffer (150 mM KAc, pH 5.5; 140 mM NaCl; 5 mM EDTA; 0.5 mM DTE; 0.5 mM PMSF).

3. Dilute the snRNP preparation to 133 mg/ml with 4 M KAc, pH 5.5; 4.0 M NaCl; 500 mM EDTA; 500 mM DTE; 500 mM PMSF and sterile water so that the mixture contains 150 mM KAc; 140 mM NaCl; 5 mM EDTA; 0.5 mM DTE; 0.5 mM PMSF.

4. Add the diluted snRNP preparation to the DE53 resin into a test tube.

5. Incubate the mixture for 15 min on ice followed by 15 min at 37 °C, keeping the resin in suspension by inverting the tube once per min.

6. Centrifuge the mixture at 16 000 g for 10 min at 4 °C in the Sorvall HB4 rotor. Remove the supernatant and store it on ice.

7. Resuspend the DE53 resin in a onefold volume of wash buffer, and incubate it for 15 min at 37 °C, mixing by inverting once per min.

8. Centrifuge the resin and remove the supernatant as in step 6.

9. Combine the two supernatants (from steps 6 and 8) and dialyze (using a membrane with a 3.5 kDa exclusion) 2 h at 4° C against a 50-fold volume of reconstitution buffer (20 mM Hepes-KOH, pH 7.9; 50 mM KCl; 5 mM $MgCl_2$; 0.2 mM EDTA; 5 % glycerol; 0.5 mM DTE; 0.5 mM PMSF).

10. Dialyze the preparation at 4° C against a 30-fold volume of reconstitution buffer containing 30 % PEG 6000 until a 50- to 100-fold reduction in volume (with a protein concentration of at least 200 mg/ml) is achieved (typically 4–6 h).

11. Dialyze the concentrated preparation for 15 min at 4 °C against reconstitution buffer.

12. Determine the protein concentration (Bearden 1978) and analyze both the protein and RNA by PAGE (Will et al. 1993).

Reconstitution of snRNP core particles

1. Mix 5 µg U1 snRNA, 100 µg purified snRNP protein; 1.0 M Hepes-KOH, pH 7.9; 2.0 M KCl; 500 mM $MgCl_2$; 100 mM DTE; RNasin in a final volume of 0.2–0.5 ml, such that the final reconstitution mixture contains 20 mM Hepes-KOH, pH 7.9; 50 mM KCl; 5 mM $MgCl_2$, 1 mM DTE; and 0.5 units/ml of RNasin. Ensure that the protein concentration is at least 100 mg/ml.

2. Incubate the mixture for 30 min at 30° C, followed by 15 min at 37 °C.

3. Separate the reconstituted snRNPs from free proteins and RNA by glycerol or sucrose gradient centrifugation.

5.2
Electron Microscopy of snRNPs

The pioneering studies of ribosomal subunit structure by Stöffler and Lake and their colleagues (Tischendorf et al. 1974a,b, 1975; Lake 1976, 1978; Stöffler and Stöffler-Meilicke 1984) demonstrated the power of negatively stained samples in EM for determining the structure including the spatial arrangement of the individual components of RNP particles. The snRNPs, which sediment with coefficients between 10S and 25S, are nearly similar in size to ribosomal subunits and thus can be readily investigated by classical EM methods. Indeed, significant information regarding the higher order structure of the spliceosomal snRNPs has been obtained by EM. The individual snRNP particles have been shown to be structurally asymmetric and can be examined by immuno-EM, in which antibodies are used to mark the position of the individual snRNP components. In this way, it is possible to localize the relative positions of individual snRNP proteins, snRNA sequences, cofactors and even the interaction site of a given snRNP with pre-mRNA or other snRNPs. For the visualization of snRNPs, negative staining by the double carbon film method, described by Tischendorf et al. (1974a), is recommended. In this procedure, the RNP particles are first adsorbed to a carbon film, contrasted with uranyl formate, and then sandwiched between a second carbon film. The EM imaging of negatively contrasted snRNPs allows clear visualization of the particle's outlines, but internally located fine structures are less well contrasted. The lack of internal resolution is likely caused by a specimen-flattening effect due to the cohesive forces present in the carbon sandwich (Lake 1976). Nonetheless, this effect appears to enhance the imaging of snRNP-antibodies complexes, since IgGs are forced into a coplanar orientation, improving the recognition of their characteristic Y-shape.

Negative Staining by the Carbon Double Film Method

As described above, the negative staining of snRNPs is carried out with thin carbon films which serve as snRNP sample carriers. The carbon films best suited for snRNP adsorption are generated by the indirect deposition of carbon vapor on freshly

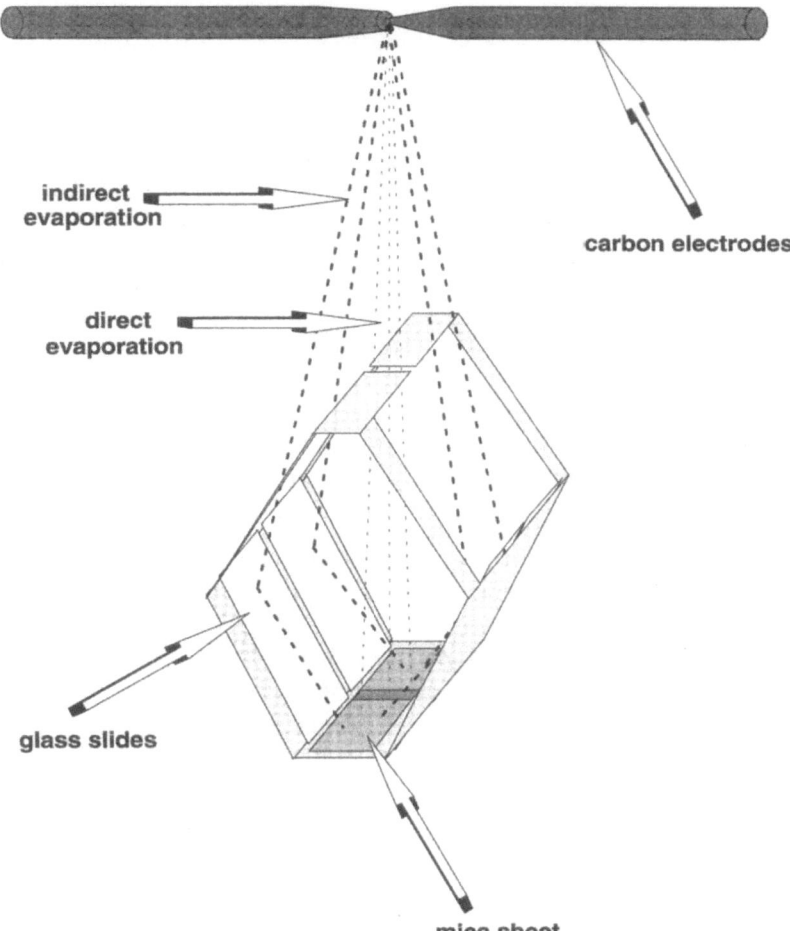

Fig. 5.9. Indirect evaporation of carbon on mica. A cache placed in the vacuum chamber holds four glass slides in a tilted upright position, and protects the mica from direct evaporation by the carbon source. The carbon emitted from the electrodes hits the mica only after reflection on the glass slides. A small area of directly evaporated carbon is helpful for visual inspection of the film

cleaved mica. The mica is positioned in a carbon vaporizing chamber, such that the carbon contacts the mica only after reflection from the surface of conventional glass slides. A possible experimental set-up is shown in Fig. 5.9. Thin films produced in this manner possess excellent staining behaviour and relatively high stability. Adsorption of the snRNP particles and subsequent staining with uranyl formate are carried out best in small holes bored into a block of black Teflon. Due to the very small sample and use of standard EM grids, with a diameter of 3.05 mm, holes with a sample capacity of approximately 45 μl are used. In order to improve the visibility of the floating carbon films generated during EM sample preparation, the Teflon holes should be filled until the solution has a completely even surface (i.e., do not over- or underfill). Visualization of the carbon films can be further improved by positioning a spot-sized illuminating beam at a low angle behind the Teflon block. The following protocol is adapted from Tischendorf et al. (1974a) and Stöffler-Meilicke and Stöffler (1988). Use reagents only of highest available purity and double distilled water. All solutions are kept at 4° C. SnRNP-containing Mono Q or gradient fractions can be used directly for EM specimen preparation. The double carbon film negative staining procedure is illustrated in Fig. 5.10.

1. Fill one hole (A in Fig. 5.10) of the Teflon block with the snRNP sample (optimal concentration is 20–30 μg/ml), and two holes (B and C) with 2.5 % uranyl formate (Polyscience) solution. The latter is light-sensitive and should be prepared directly before use by dissolving in water (mix well for 15 min) and clearing by centrifugation. Other stains could also be used in principle (see Bremer et al. 1992).

2. Using fine forceps, submerge a 3×3 mm piece of carbon-coated mica (CM) into the snRNP sample so that the greater part of the carbon film detaches and floats to the surface (see panel 1 in Fig. 5.10). Allow the snRNPs to adsorb to the film for 20 s to 20 min. The exact incubation time depends upon several variables, including the snRNP concentration, and must be empirically determined.

3. Remove the mica with the carbon film from the solution (panel 2 in Fig. 5.10), drain excess of sample solution, and transfer to the 2.5 % uranyl formate solution, allowing the carbon film to completely detach from the mica (panels 3, 4 in Fig. 5.10).

4. After 3 min remove the carbon film by placing an EM grid containing a perforated carbon film on the sample side (G in Fig. 5.10) directly on top of it and lifting (panels 4, 5 in Fig. 5.10). The preparation of grids layered with a perforated supporting carbon film is described in Lünsdorf and Spiess (1986), Jahn (1995) and Fukami et al. (1965).

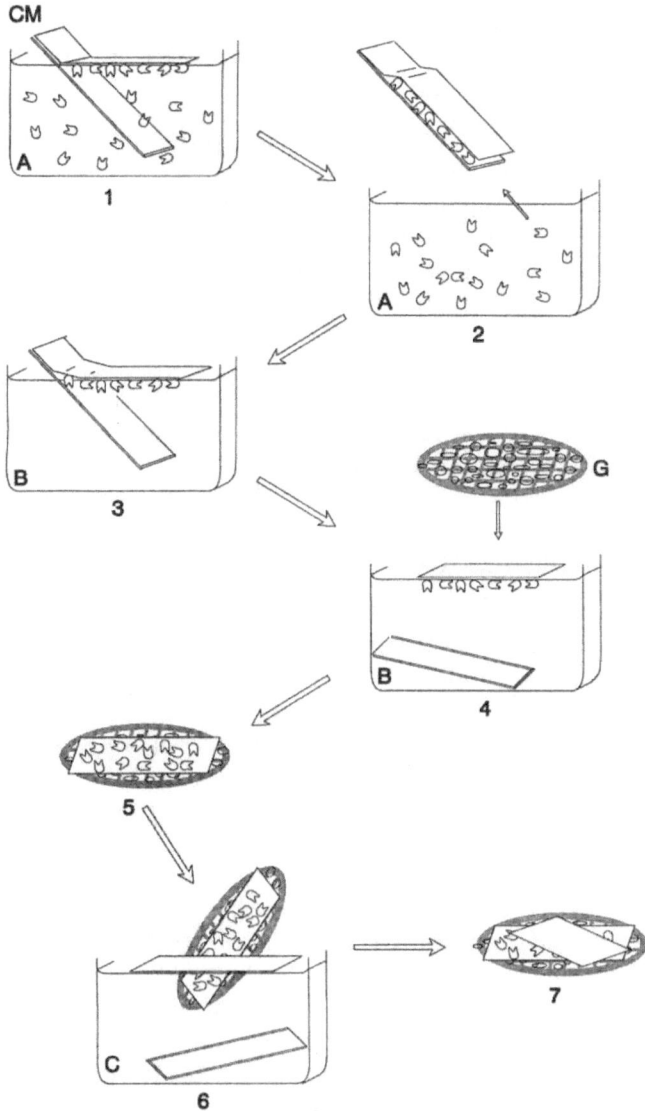

Fig. 5.10. The double carbon film negative staining procedure. *A*, *B*, and *C* indicate the three small holes in a black Teflon block. *A* is filled with the snRNP sample, while *B* and *C* are filled with 2.5 % uranyl format solution. *CM* carbon-coated mica; *G* EM grid covered by a perforated carbon film. See text for the procedure description

5. Submerge a second carbon-coated mica plate in 2.5 % uranyl formate, allowing the carbon film to completely detach (panel 6 in Fig. 5.10).

6. Submerge the grid from step 5 (containing the attached carbon film with adsorbed snRNPs; 5 in Fig. 5.10) underneath the second floating carbon film, and lift out of the solution with the snRNP side up, such that the snRNPs are trapped between two carbon film layers (panels 6, 7 in Fig. 5.10). Remove excess stain by touching the grid edges with a filter paper.

7. Allow to air dry.

8. Image with a transmission EM operating at 80 kV acceleration voltage. Electron micrographs are prepared at a primary magnification of 60 000–80 000 for large snRNPs or up to 100 000 for small snRNPs. For subsequent examination, they are photographically enlarged to a final magnification of 300 000–500 000.

Electron micrographs of 12S U1 snRNPs and 20S US snRNPs, prepared as described in the protocol, are shown in Fig. 5.11. Negatively stained U1 snRNPs (Fig. 5.11A) possess an almost circular main body, ca. 8 nm in diameter, with two characteristic protuberances, 4–7 nm long and 3–4 nm wide. The 20S U5 snRNP (Fig. 5.11B) appears as an elongated structure, 20–23 nm in length and 11–14 nm wide. The segmentation line dividing the particle into a head and a body region can be seen; the latter, in most cases, appears to be pointed due to the presence of one or two short protuberances. The apparent variations in the observed form of a given snRNP result mainly from different particle orientations on the carbon film; that is, they arise from different two-dimensional (2D) projections of a three-dimensional (3D) object. Other factor, such as biochemical heterogeneity, conformational flexibility, beam damage, background noise and variations in the staining effect influence the appearance of the particles on the micrographs as well.

In this chapter, interpretation of the EM images are performed by visual inspection and classification. For a more sophisticated interpretation of the images computer-aided image analysis can be carried out. With the methods developed by van Heel and Frank (1981) images of negatively stained ribosomes have been classified and class members finally averaged to reduce noise and to enhance resolution of characteristic views (reviewed in Frank et al. 1988; Harauz et al. 1988).

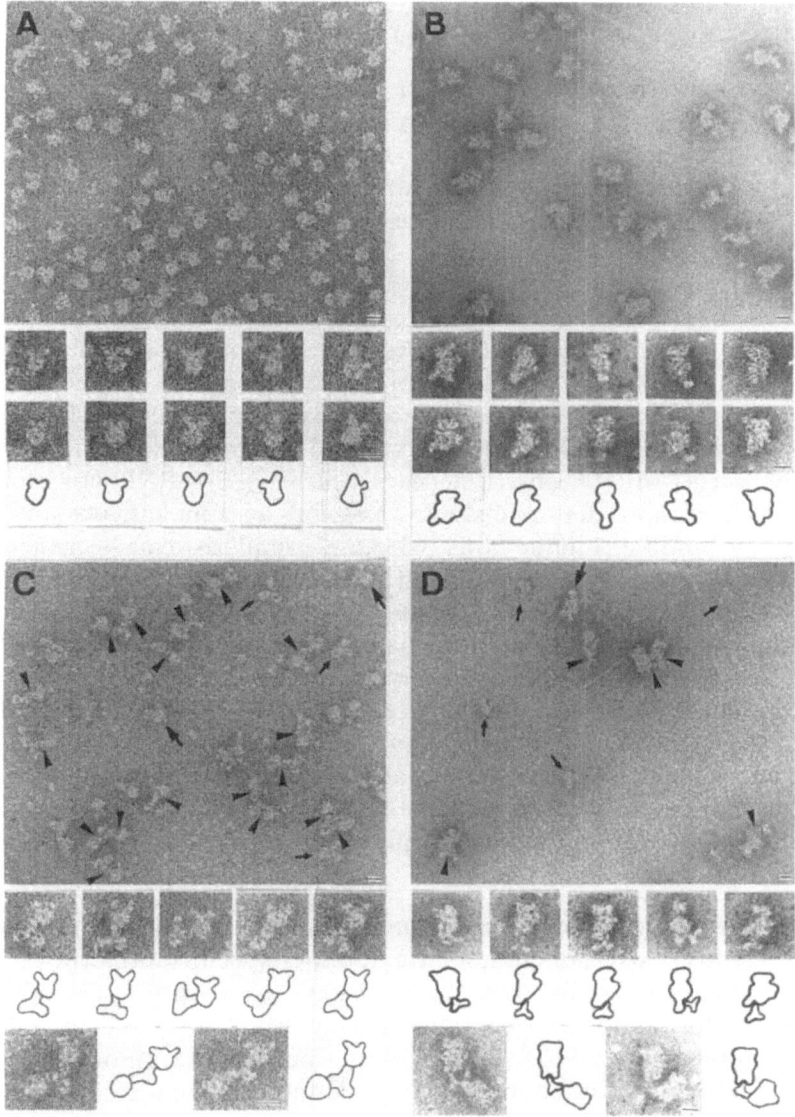

Fig. 5.11A–D. Overviews and selected micrographs of negatively stained snRNPs and immunocomplexes. U1 snRNPs (**A**), 20S U5 snRNPs (**B**) and immune complexes of IgG antibodies specific for the m₃G cap bound to either U1 snRNPs (**C**) or 20S U5 snRNPs (**D**) are shown

Specific Labeling of Proteins at snRNP Particles

As previously mentioned, EM specimen preparation by the double carbon film method is especially well-suited for electron microscopic imaging of snRNP-IgG complexes. IgGs are preferably used for labeling studies since their characteristic Y-shape can generally be easily distinguished. The Fab arm of the IgG can be readily resolved by EM and, thus, the position of the antigen determinant can be localized through its interaction with the Fab arm. The maximum resolution with which an antibody binding site can be localized is limited by the size of the Fab arm, which is 3–3.5 nm. Due to the presence of two Fab arms per IgG molecule, two types of immunocomplexes can be formed. If only one antibody binding site is accessible on a particle, then a binary IgG-snRNP and a ternary snRNP-IgG-snRNP complex result. If more than one site is accessible for simultaneous antibody binding, higher order immune complexes can be generated. By labeling a particle with two antibodies of different specificities simultaneously, distances between the two antigenic sites on the particle can be determined (Kastner et al. 1981, 1992; Lake 1982). Although both polyclonal and monoclonal antibodies have been successfully used for investigating the higher order structure of ribosomal subunits, monoclonal or affinity-purified polyclonal antibodies are recommend for immuno-EM investigations (see also Lake 1978; Spiess et al. 1987; Glitz et al. 1988; Stöffler-Meilicke and Stöffler 1988; Boublik 1990).

Immuno-globulin-snRNP complex formation

1. Immunocomplexes are generated by incubating purified snRNPs with an equimolar amount of specific antibody. The length of incubation is highly dependent upon the reactivity of the antibody, generally varying from 1 to 24 h at 4° C (Spiess et al. 1987). To control for nonspecific antibody interactions, it is advisable to analyze immune complexes formed, either in the presence of excess isolated antigen or snRNP particles which lack the target antigen.

2. EM preparations can be made directly from the antibody-snRNP reaction mixture, provided relatively efficient complex formation is observed. Otherwise, the free antibodies can be separated from unbound snRNPs by centrifugation through linear sucrose gradients. In order to minimize sample input and to prevent significant dilution of the snRNP-IgG complexes, the use of 1.5 ml gradients with the TLS-55 swinging

bucket rotor is recommended. The small (12S) snRNPs are fractionated by centrifugation at 55 000 rpm for 5 h at 4° C on 5 %–20 % (w/w) sucrose gradients prepared in buffer A (25 mM Hepes-KOH, pH 7.9; 150 mM KCl; 1.5 mM MgCl$_2$). Larger particles (17–25S) can be separated from IgGs on 10–30 % (v/v) glycerol gradients prepared in buffer A by centrifugation at 55 000 rpm for 2–3 h at 4 °C. Note that the centrifugation times used here are shorter than those given in Table 5.1 to prevent pelleting of the larger immune complexes.

3. Measure the optical density (at 254 or 260 nm), either during automatic fractionation or manually for each fraction, to create an snRNP sedimentation profile. The efficiency of snRNP-IgG complex formation can be estimated from this profile by comparing the area of the free snRNP peak with that of the faster sedimenting snRNP-IgG complexes.

4. To quantitatively determine the amounts of free IgG and snRNPs as well as IgG-snRNP complexes in each fraction, a microtiter ELISA can be performed as described in the following protocol.

To obtain an optimal yield of immunocomplexes, the appropriate antibody concentration for incubation with the snRNPs has to be determined. Gradient centrifugation of the antibody-snRNP incubation mixture allows the formation effeciency of ternary immunocomplexes (i.e., with two snRNPs per IgG molecule) to be determined by monitoring the optical density during fractionation. Ternary immunocomplexes sediment significantly faster than free snRNPs and can be thus identified in the UV adsorption peaks at high S values (see also Stöffler-Meilicke and Stöffler 1988). Binary immunocomplexes, by contrast, cannot be identified by the optical density profile as these often cosediment with the antibody-free snRNP particles. To determine the snRNP and antibody concentrations independent of each other, and thus the formation effeciency of binary immune complexes, an ELISA assay of the gradient fractions can be performed. This information can be important when only binary immune complexes are formed or when ternary immune complexes are not desired (e.g., if their images are too difficult to interpret). Here, the detection of U1 snRNP and a monoclonal mouse IgG antibody is described (Kastner and Lührmann 1989). However, the protocol can be modified for the detection of different snRNP particles or other types of monospecific antibodies.

Analysis by microtiter ELISA

1. Pipet 1/4 of each gradient fraction (up to 50 µl) into two separate ELISA microtiter plates and incubate either at 4 °C overnight or at room temperature 2 h. The final sample volume in each well should be 50 µl (add PBS if necessary).

2. Wash the cavities once with PBS, and saturate the nonspecific binding sites with 150 µl 1 % BSA (RIA grade) in PBS by incubating overnight at 4 °C or for 3 h at room temperature.

3. Wash three times with PBS.

4. To determine the snRNP concentration, pipet 80 µl of anti-RNP serum, diluted 1:1000 in ELISA buffer (PBS pH 7.4, 0.1 % [v/v] Tween-20) containing 1 % BSA, into the wells of plate A, and incubate overnight at 4° C or 2 h at room temperature.

5. Wash plate A three times with ELISA buffer and incubate 1 h at 4 °C with 80 µl phosphatase-conjugated anti-human-IgG antibodies (diluted 1:1000 in ELISA buffer).

6. To determine the IgG concentration, pipet 80 µl phosphatase-conjugated anti-mouse-IgG antibody, diluted 1:1000 in ELISA buffer, into the wells of plate B and incubate 1 h at 4 °C.

7. Wash both microtiter plates three times with PBS.

8. Add 80 µl freshly prepared substrate buffer (1 mg/ml p-nitrophenyl phosphate; 100 mM sodium carbonate, pH 9.5; 2 mM $MgCl_2$) and incubate at room temperature for 30 min – 4 h. During this incubation, measure the absorption at 450 and 405 nm with an ELISA reader. Determine the relative phosphatase activity bound to the plate by subtracting of the absorption value at 450 nm from the 405 nm value.

9. Plot the activity values over the fraction numbers and estimate the relative amounts of free IgG and snRNPs, and IgG-snRNP complexes.

Gradient centrifugation can be used not only to monitor optimal conditions for IgG-snRNP interaction but also to determine whether sufficiently stable complexes have been obtained, since IgG-snRNP complexes which withstand gradient centrifugation generally do not dissociate during the EM sample preparation. To prevent the dissociation of snRNP proteins only weakly associated, it may be necessary to perform chemical cross-linking prior to the addition of antibody. The reagent dithio-bis-succinimidyl-propionate (DSP), which cross-links primary and

secondary amino groups, has been successfully used to covalently attach Ul snRNP proteins without significantly altering the
shape of the particle (Kastner et al. 1992). The chemical cross-
linking procedure described in the protocol is adapted from
Lomant and Fairbanks (1976).

1. Dialyze snRNPs for 3 h at 4 °C against a 50-fold volume of
   cross-linking buffer (20 mM triethanolamine, pH 8.5; 300 mM
   KCl; 1.5 mM MgCl$_2$).

2. Slowly add DSP (Pierce) solution (18 mM in DMSO) to the
   snRNP sample (150–300 µg protein/ml) until a final concentration of 60 µM is reached. This DSP concentration was
   found to be the best compromise between stabilization of the
   U1 snRNP particle and retaining antibody reactivity with
   anti-70K and anti-A antibodies (Kastner et al. 1992). For
   other proteins, the optimal DSP concentration may differ.

3. Incubate at 0 °C for 30 min.

4. Terminate the reaction by adding glycinamide hydrochloride
   (3 M, pH 4.0) to a final concentration of 50 mM; incubate at
   37 °C for 40 min.

5. Cross-link formation can be monitored by PAGE of the
   snRNP proteins analyzed and extracted with and without
   DTE or mercaptoethanol. The cross-linked particles (from
   step 4) can be used directly for immuno-snRNP complex
   formation or EM sample preparation. The extent of inter-
   snRNP cross-linking can be analyzed by sucrose gradient
   centrifugation. By ELISA analysis, the effect of cross-linking
   on the antibody reactivity can be monitored.

**snRNP Pro-
tein Cross-
Linking with
DSP**

Localization of snRNA Sequences by Oligonucleotide Labeling

The prerequisite for using antibody labeling to localize RNA
sites is the availability of an antibody which specifically reacts
with an RNA site. In most cases, RNA-specific antibodies are difficult to produce. Antibodies which specifically bind modified
nucleosides, such as the anti-m$_3$G antibody, might be available.
Antibodies directed against a particular snRNA sequence have
been isolated from autoimmune sera. Using an autoantibody
specific for the furthest 3' stem/loop of U1 RNA, this RNA
sequence could be located on the U1 snRNP using immunoelectron microscopy (Hoet et al. 1993).

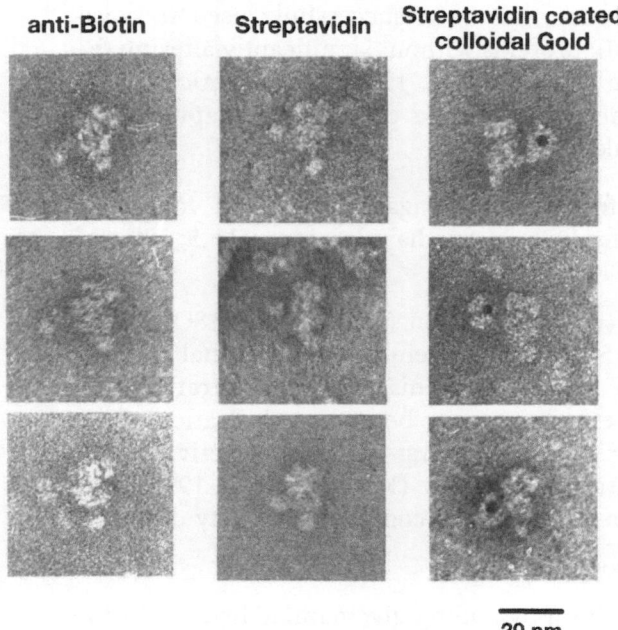

anti-Biotin   Streptavidin   Streptavidin coated colloidal Gold

20 nm

Fig. 5.12. U5 snRNPs with a label bound to loop I of U5 RNA. 20S U5 snRNPs were labeled with a loop I-specific oligonucleotide which was attached to an anti-biotin IgG antibody, streptavidin, or streptavidin-coated colloidal gold. The labels are always located on the central main body of the 20S U5 snRNP

A more generalized approach for the localization of accessible RNA sequences is labeling with complementary oligonucleotides. This method was first developed for locating rRNA sequences at the surface of ribosomal subunits (Oakes et al. 1986; Olsen et al. 1988). This method also proved useful for labeling snRNPs (Kastner et al. 1992) and pre-spliceosomal complexes (Furman and Glitz 1995). Compared to ribosomes, the snRNPs have much less RNA exposed to the solvent, so that many RNA sequences cannot be targeted by this method. However, the scarcity of RNA sites available can be advantageous for the oligonucleotide binding specificity. A prerequisite for the oligonucleotide labeling method is the availability of an oligonucleotide (DNA, RNA or 2'-O-modified RNA) that binds stably and specifically to the target RNA sequence within the RNP particle. In addition, the oligonucleotide must be bound to a marker which can be visualized and recognized in the EM when present in a complex with the labeled snRNP. Usually the EM-visible marker is bound to the oligonucleotide via a hapten molecule, which is covalently attached to the oligonucleotide. Most frequently, biotin is used as the hapten, and streptavidin or biotin-specific antibodies as EM-visible markers (Glitz et al. 1988; Scheinman et al. 1992).

To form labeled snRNPs, the oligonucleotide is initially bound to the snRNP, and then the marker is attached to the snRNP-bound oligonucleotide. Alternatively, a complex of marker and oligonucleotide can be formed prior to binding with snRNP. For the labeling of the U5 RNA loop I at the 20S U5 snRNP, the latter method has been shown to be most effective, with the markeroligonucleotide complex purified by size exclusion chromatography before incubation with the snRNP (Kastner et al. unpublished).

In order to label loop I at the 20S U5 snRNP, a biotinylated, base-modified, 2'-O-allyloligoribonucleotide was tested with three kinds of biotin-binding markers: streptavidin, streptavidin-coated colloidal gold, and a biotin-specific monoclonal IgG-antibody. The best result was obtained with the antibody, which gave a more efficient and specific label than the others. The small size of streptavidin, in addition, makes it difficult to unambiguously identify when attached to the 20S U5 snRNP. When used to label smaller structures, like 12S U4/U6 snRNPs, a single bound streptavidin could be clearly identified (Kastner et al. 1991). The streptavidin-coated colloidal gold can be easily recognized, but its large size interferes with the localization of the binding site. In Fig. 5.12 examples of 20S U5 snRNPs labeled with each of the three markers are shown.

Biotinylated DNA oligonucleotides are suitable for many labeling tasks. When particular stabile hybrids between the oligonucleotide and the target RNA sequence are necessary, the use of 2'-0-methyl or 2'-0-allyloligonucleotides are recommended (Lamond and Sproat 1993). These oligonucleotides bind extremely stably, so that the RNA hybrids cannot be dissociated without snRNP denaturation. However, the high price of synthons (Boehringer Mannheim) and of commercial synthesis of 2'-0-methyl or 2'-0-allyloligonucleotides (which is ten times more than DNA oligonucleotides) may be a limiting factor in their use.

1. Synthesize a 3' or 5' biotinylated, 2'-O-methyl or 2'-O-allyloligonucleotide complementary to the target RNA sequence. The oligo can also be ^{32}P-labeled for monitoring and quantification of subsequent binding step efficiencies.

Formation of antibody-tagged snRNP-oligo-nucleotide complexes

2. Incubate 25 µg purified monoclonal, anti-biotin IgG antibody (Sigma) with stoichiometric amounts of the oligonucleotide in PBS (pH 8.0) for 1 h.

3. Separate free oligonucleotides from the antibodies and antibody-oligonucleotide complexes by gel filtration (LKB Ultrogel AcA).

4. Incubate the fractions containing the oligonucleotide-antibody complex in buffer P (20 mM Tris-HCl, pH 7.9; 150 mM KCl; 5 mM $MgCl_2$) with 20 pmol snRNPs for 30 min at 30 °C and then 60 min at 0 °C. Use an oligonucleotide lacking complementarity to the snRNA as a negative control.

5. Separate free antibodies and oligonucleotides by sucrose gradient centrifugation in a TLS 55 rotor (see Table 5.1).

6. Analyze the snRNP fractions by EM and determine the specific-antibody binding site.

The strategy of labeling via antibody binding to a hapten molecule was also used for the localization of rRNA, tRNA, mRNA, inhibitors (reviewed in Stöffler and Stöffler-Meilicke 1984) and proteins (Bergmann and Wittmann-Liebold 1990; Montesano-Roditis et al. 1993) on ribosomal subunits.

Localization of Specific Sites by EM

The micrographs of negatively stained specimens represent 2D projections of the imaged 3D object. The 2D projections provide information on the general shape of the object, and a bound label can be identified and localized if it has a distinct appearance. The interpretation of immunocomplex images generated by EM has been described in detail for ribosomal subunits by Stöffler-Meilicke and Stöffler (1988); this description also applies here. In Fig. 5.11 (C and D), examples of U1- or U5 snRNP-anti-m_3G antibody complexes are shown. For U1 snRNPs, the EM samples were prepared directly from the IgG-snRNP reaction mixture, whereas U5 snRNP-IgG complexes were partially purified by gradient centrifugation. The binding site of the anti-m_3G antibody (and hence the position of the snRNA m_3G cap) can be determined directly from the 2D projections. These are located on the bottom of the U5 snRNP body, close to the base of the lower protuberance, and on the U1 snRNP body, close to one protuberance and approximately opposite the other. Recognition of the anti-m_3G IgG binding site on the U1 particle requires practice, since the Ul snRNP and IgG possess a similar general shape and size. The two components can be distinguished, however, since the U1 snRNP protuberances are somewhat thinner than the antibody arms. For further 3D localization, reliable 3D models of the snRNPs are needed. Negatively stained specimens are not the ideal objects for 3D reconstruction because of preferential staining due to inhomogeneous stain affinities of the particle surface and to staining polarities introduced by the supporting carbon film. In this case, other EM methods employing unstained specimens are more suitable. The recent developments in low dose imaging and sophisticated 3D reconstruction algorithms have made 3D EM successful, even from frozen hydrated aperiodic samples (Frank 1995; Frank et al. 1995; Stark et al. 1995, Skoglund et al. 1996).

Acknowledgements. I would like to thank R. Lührmann for his continuous support and collaboration as well as the members of the Lührmann laboratory for their contributions to this manuscript, in particular C.L. Will and V.A. Raker for critically reading the manuscript. I also like to thank the EM laboratory of R. Lurz at the Max-Planck-Institut für Molekulare Genetik in Berlin for many helpful advices. This work was supported by grants from the Deutsche Forschungsgemeinschaft (Ka 805/2–1 to B.K. and SFB 272 to Reinhard Lührmann).

References

Ast G, Weiner AM (1996) A U1/U4/U5 snRNP complex induced by a 2'-O-methyl-oliogoribonucleotide complementary to U5 snRNA. Science 272:881–884

Bach M, Winkelmann G, Lührmann R (1989) 20S small nuclear ribonucleoprotein U5 shows a surprisingly complex protein composition. Proc Natl Acad Sci USA 86:6038–6042

Bach M, Bringmann P, Lührmann R (1990a) Purification of small nuclear ribonucleoprotein particles with antibodies against modified nucleosides of small nuclear RNAs. Methods Enzymol 181:232–257

Bach M, Krol A, Lührmann R (1990b) Structure-probing of U1 snRNPs gradually depleted of the U1-specific proteins A, C and 70k. Evidence that A interacts differentially with developmentally regulated mouse U1 snRNA variants. Nucleic Acids Res 18:449–457

Bearden JC Jr (1978) Quantitation of submicrogramm quantities of protein by an improved protein-dye binding assay. Biochim Biophys Acta 533:525–529

Behrens SE, Tyc K, Kastner B, Reichelt J, Lührmann R (1993) Small nuclear ribonucleoprotein (RNP) U2 contains numerous additional proteins and has a bipartite RNP structure under splicing conditions. Mol Cell Biol 13:07–319

Behrens SE, Kastner B, Lührmann R (1994) Preparation of U small nuclear ribonucleoprotein particles In: Celis JE (ed) Cell biology – a laboratory handbook, vol 1. Academic Press, London, pp 628–639

Behrens S-E, Lührmann R(1991) Immunoaffinity purification of a [U4/U6U5] tri-snRNP from human cells. Genes Dev 5:1439–1452

Bennett M, Michaud S, Kingston J, Reed R (1992) Protein components specifically associated with prespliceosome and spliceosome complexes. Genes Dev 6:1986–2000

Bergmann U, Wittmann-Liebold B (1990) Use of a hapten specific anti-dansyl antibody for the localization of ribosomal proteins by immuno electron microscopy. Biochem Int 21:741–751

Bochnig P, Reuter R, Bringmann P, Lührmann R (1987) A monoclonal antibody against 2,2,7-trimethylguanosine that reacts with intact U_1 snRNPs as well as with 7-methylguanosine-capped RNAs. Eur J Biochem 168:461–467

Boublik M (1990) Electron microscopy of ribosomes In: Spedding G (ed) Ribosomes and protein synthesis – a practical approach. IRL, Oxford, pp 273–296

Bremer A, Henn C, Engel A, Baumeister W, Aebi U (1992) Has negative staining still a place in biomacromolecular electron microscopy? Ultramicroscopy 46:85–111

Bringmann P, Rinke J, Appel B, Reuter R, Lührmann R (1983) Purification of snRNPs U1, U2, U4, U5 and U6 with 2,2,7-trimethylguanosine-specific antibody and definition of their constituent proteins reacting with anti-Sm and anti-(U1)RNP antisera. EMBO J 2:1129–1135

Bringmann P, Lührmann R (1986) Purification of the individual snRNPs U1, U2, U5 and U4/U6 from HeLa cells and characterization of their protein constituents. EMBO J 5:3509–3516

Brosi R, Gröning K, Behrens S-E, Lührmann R, Krämer A (1993a) Interaction of mammalian splicing factor SF3a with U2 snRNP and relation of its 60-kD subunit to yeast PRP9. Science 262:102–105

Brosi R, Hauri H-P, Krämer A (1993b) Separation of splicing factor SF3 into two components and purification of SF3a activity. J Biol Chem 268:17640–17646

Dignam JD, Lebovitz RM, Roeder RG (1983) Accurate transcription initiation by RNA polymerase II in a soluble extract from isolated mammalian nuclei. Nucleic Acids Res 11:1475–1489

Fabrizio P, Esser S, Kastner B, Lührmann R (1994) Isolation of S. cerevisiae snRNPs:comparison of U1 and U4/U6U5 to their human counterparts. Science 264:261–265

Frank J, Radermacher M, Wagenknecht T, Verschoor A (1988) Studying ribosome structure by electron microscopy and computer-image processing. In: Noller HF, Moldave K (eds) Methods in enzymology, vol 164. Academic Press, San Diego, pp 3–35

Frank, J (1995) Approaches to large-scale structures. Curr Opin Struct Biol 5:194–201

Frank J, Zhu J, Penczek P, Li Y, Srivastava S, Verschoor A, Radermacher M, Grassucci R, Lata RK, Agrawal RK (1995) A model of protein synthesis based on cryo-electron microscopy of the *E. coli* ribosome. Nature 376:441–444

Frendeway D, Keller W (1985) Stepwise assembly of a pre-mRNA splicing complex requires U-snRNPs and specific intron sequences. Cell 42:355–367

Fukami A, Adachi K (1965) A new method of preparation of a self-perforated micro-plastic grid and its application. J Elec Microsc (Tokyo) 14:112–118

Furman E, Glitz DG (1995) Purification of the spliceosome A-complex and its visualization by electron microscopy. J Biol Chem 270:15515–15522

Glitz DG, Cann PA, Lasater LS, Olson HM (1988) Antibody probes of ribosomal RNA. Methods Enzymol 164:493–502

Grabowski PJ, Seiler SR, Sharp PA (1985) A multi-component complex is involved in the splicing of messengerRNA precursor. Cell 42:345–353

Gröning K, Krämer A, Kastner B (1997) (in preparation)

Guthrie C, Patterson B (1988) Spliceosomal snRNAs. Annu Rev Genet 22:387–419

Harauz G, Boekema E, Van Heel M (1988) Statistical image analysis of electron micrographs of ribosomal subunits. In: Noller HF, Moldave K (eds) Methods in enzymology, vol 164. Academic Press, San Diego, pp 35–49

Hoet RM, Kastner B, Lührmann R, Van Venrooij WJ (1993) Purification and characterization of human autoantibodies directed to specific regions on U1RNA; recognition of native U1RNP complexes. Nucleic Acids Res 21:5130–5136

Jahn W (1995) Easily prepared holey films for use in cyro-electron microscopy. J Microsc 179:333–334

Kastner B, Stöffler-Meilicke M, Stöffler G (1981) Arrangement of the subunits in the ribosome of *Escherichia coli*: demonstration by immuno-electron microscopy. Proc Natl Acad Sci USA 78:6652–6656

Kastner B, Bach M, Lührmann R (1990) Electron microscopy of small nuclear ribonucleoprotein (snRNP) particles U2 and U5: evidence for a common structure-determining principle in the major U snRNP family. Proc Natl Acad Sci USA 87:1710–1714

Kastner B, Bach M, Lührmann R (1991) Electron microscopy of U4/U6 snRNP reveals a Y-shaped U4 and U6 RNA containing domain protruding from the U4 core RNP. J Cell Biol 112:1065–1072

Kastner B, Kornstädt U, Bach M, Lührmann R (1992) Structure of the small nuclear RNP particle U1: identification of the two structural protuberances with RNP-antigens A and 70K. J Cell Biol 116:839–849

Kastner B, Lührmann R (1989) Electron microscopy of U1 small nuclear ribonucleoprotein particles: Shape of the particle and position of the 5' RNA terminus. EMBO J 8:277–286

Krämer A (1990) Purification of small nuclear ribonucleoprotein particles active in RNA processing. Methods Enzymol 181:215–231

Krämer A (1992) Purification of splicing factor SF1, a heat-stable protein that functions in the assembly of a presplicing complex. Mol Cell Biol 12:4545–4552

Krämer A, Utans U (1991) Three protein factors (SF1, SF3 and U2AF) function in pre-splicing complex formation in addition to snRNPs. EMBO J 10:1503–1509

Lake JA (1976) Ribosome structure determined by electron microscopy of *Escherichia coli* small subunits, large subunits and monomeric ribosomes. J Mol Biol 105:131–139

Lake JA (1978) Electron microscopy of specific proteins. Three dimensional mapping of ribosomal proteins using antibody labels. In: Koehler JK (ed) Advanced techniques in biological electron microscopy II. Springer, Berlin Heidelberg New York, pp 173–211

Lake JA (1982) Ribosomal subunit orientations determined in the monomeric ribosome by single and by double-labeling immune electron microscopy. J Mol Biol 161:89–106

Lamond AI, Sproat BS (1993) Isolation and characterization of ribonucleoprotein complexes In: Higgins SJ, Hames BD (eds) RNA processing – a practical approach, vol I. IRL, Oxford, pp 103–140

Lomant AJ, Fairbanks G (1976) Chemical probes of extended biological structures: synthesis and properties of the cleavable protein cross-linking reagent [35S] dithiobis(succinimidyl propionat). J Mol Biol 104:243–261

Lührmann R, Appel B, Rinke J, Reuter R, Rothe S, Bald R (1982) Isolation and characterization of rabbit anti-m^{2,2,7} G antibodies. Nucleic Acids Res 10:7103–7113

Lührmann R, Kastner B, Bach M (1990) Structure of spliceosomal snRNPs and their role in pre-mRNA splicing. Biochim Biophys Acta 1087:265–292

Lünsdorf H, Spiess E (1986) A rapid method of preparing perforated supporting foils for thin carbon films used in high resolution transmission electron microscopy. J Microsc 144:211–213

Madhani HD, Guthrie C (1994) Dynamic RNA-RNA interactions in the spliceosome. Annu Rev Genet 28:1–26

Miriami E, Angenitzki M, Sperling R, Sperling J (1995) Magnesium cations are required for the association of U small nuclear ribonucleoproteins and SR proteins with pre- mRNA in 200 S large nuclear ribonucleoprotein particles. J Mol Biol 246:254–263

Montesano-Roditis L, McWilliams R, Glitz DG, Olah TV, Perrault AR, Cooperman BS (1993) Placement of dinitrophenyl-modified ribosomal proteins in totally reconstituted Escherichia coli 30 S subunits. Localization of proteins S6, S13, S16, and S18 by immune electron microscopy. J Biol Chem 268:18701–18709

Moore MJ, Query CC, Sharp PA (1993) Splicing of precursors to mRNA by the spliceosome In: Gesteland RF, Atkins JF (eds) The RNA world. Cold Spring Harbor Lab, Cold Spring Harbor, New York, pp 303–357

Newman A (1994) Small nuclear RNAs and pre-mRNA splicing. Curr Opin Cell Biol 6:360–367

Nilsen TW (1994) RNA-RNA interactions in the spliceosome: Unraveling the ties that bind. Cell 78:1–4

Oakes MI, Clark MW, Henderson E, Lake JA (1986) DNA hybridization electron microscopy: ribosomal RNA nucleotides 1392–1407 are exposed in the cleft of the small subunit. Proc Natl Acad Sci USA 83:275–279

Olson HM, Lasater LS, Cann PA, Glitz DG (1988) Messenger RNA orientation on the ribosome. Placement by electron microscopy of antibody-complementary oligodeoxynucleotide complexes. J Biol Chem 263:15196–15204

Plessel G, Lührmann R, Kastner B (1997) Electron microscopy of assembly intermediates of the snRNP (Sm) core: morphological similarities between the RNA-free [EFG] protein heteromer and the intact snRNP core. J Mol Biol 265:87–94

Raker VA, Plessel G, Lührmann R (1996) The snRNP core assembly pathway: identification of stable core protein heteromeric complexes and an snRNP subcore particle in vitro. EMBO J 15:2256–2269

Reed R, Griffith J, Maniatis T (1988) Purification and visualization of native spliceosomes. Cell 53:949–961

Scheinman A, Atha T, Aguinaldo AM, Kahan L, Shankweiler G, Lake JA (1992) Mapping the three-dimensional locations of ribosomal RNA and proteins. Biochimie 74:307–317

Scherly D, Boelens W, Dathan NA, Van Venrooij WJ, Mattaj IW (1990) Major determinants of the specificity of interaction between small nuclear ribonucleoproteins U1A and U2B" and their cognate RNAs. Nature 345:502–506

Ségault V, Will CL, Sproat BS, Lührmann R (1995) In vitro reconstitution of mammalian U2 and U5 snRNPs active in splicing: Sm proteins are functionally interchangeable and are essential for the formation of functional U2 and U5 snRNPs. EMBO J 14:4010–4021

Skoglund U, Öfverstedt L-G, Burnett RM, Bricogne G (1996) Maximum-entropy three-dimensional reconstruction with deconvolution of the contrast transfer function: a test application with adenovirus. J Struct Biol 117:173–188

Spann P, Feinerman M, Sperling J, Sperling R (1989) Isolation and visualization of large compact ribonucleoprotein particles of specific nuclear RNAs. Proc Natl Acad Sci USA 86:466–470

Spiess M, Zimmermann H-P, Lünsdorf H (1987) Negative staining of protein molecules and filaments. In: Sommerville J, Scheer V (eds) Electron microscopy in molecular biology – a practical approach. IRL, Oxford, pp 147–166

Stark H, Mueller F, Orlova EV, Schatz M, Dube P, Erdemir T, Zemlin F, Brimacombe R, Van Heel M (1995) The 70S *Escherichia coli* ribosome at 23 Å resolution: fitting the ribosomal RNA. Structure 3:815–821

Stöffler G, Stöffler-Meilicke M (1984) Immunoelectron microscopy of ribosomes. Annu Rev Biophys Bioeng 13:303–330

Stöffler-Meilicke M, Stöffler G (1988) Localization of ribosomal proteins on the surface of ribosomal subunits from *Escherichia coli* using immuno-electron microscopy. In: Noller HF, Moldave K (eds) Methods in enzymology, vol 164. Academic Press, San Diego, pp 503–520

Sumpter V, Kahrs A, Fischer U, Kornstädt U, Lührmann R (1992) In vitro reconstitution of U1 and U2 snRNPs from isolated proteins and snRNA. Mol Biol Rep 16:229–240

Tischendorf GW, Zeichhardt H, Stöffler G (1974a) Determination of the location of proteins L14, L17, L18, L19, L22 and L23 on the surface of the 50S ribosomal subunit of Escherichia coli by immune electron microscopy. Mol Gen Genet 137:187–208

Tischendorf GW, Zeichhardt H, Stöffler G (1974b) Location of proteins S5, S13, S14 on the surface of the 30S ribosomal subunits from Escherichia coli as determined by immune electron microscopy. Mol Gen Genet 134:209–223

Tischendorf GW, Zeichhardt H, Stöffler G (1975) Architecture of the *Escherichia coli* ribosome as determined by immune electron microscopy. Proc Natl Acad Sci USA 72:4820–4824

Van Heel M, Frank J (1981) Use of multivariate statistical statistics in analysing the images of biological macromolecules. Ultramicroscopy 6:187–194

Walter P, Blobel G (1983) Disassembly and reconstitution of signal recognition particle. Cell 34:525–533

Will CL, Kastner B, Lührmann R (1993) Analysis of ribonucleoprotein interactions. In: Higgens SJ, Hames BD (eds) RNA processing – a practical approach, vol 1. IRL, Oxford, pp 141–177

Will CL, Fabrizio P, Lührmann R (1995) Nuclear pre-mRNA splicing. In: Eckstein F, Lilley DMJ (eds) Nucleic acids and molecular biology, vol 9. Springer, Berlin Heidelberg New York, pp 342–372

Chapter 6

Detection of Autoantibodies to Extractable Cellular Antigens

Panayiotis G. Vlachoyiannopoulos*

Introduction

When a sufficient number of antibody molecules are mixed with a soluble macromolecular antigen to which this antibody binds, large aggregates of antigen with antibody molecules occur which can be visualized in agar gel as a precipitin line (immunoprecipitation). The simplest way to indicate this reaction is to put the serum (antibody source) and the antigen source into separate wells on an agar gel and allow diffusion of antigen and antibody towards each other (Ouchterlony and Nilsson 1978). When antibody and antigen meet, they form a visible precipitin line (Ouchterlony gel immunodiffusion or Ouchterlony gel diffusion assay). This system is very valuable when: (1) the antigen is a macromolecular complex with many antibody binding sites; (2) the antigen source is a crude cellular (tissue) extract usually containing more than one antigen; (3) there is a need to examine the relatedness of unknown proteins in terms of their antigenic activity, using the patients' sera as tools to study antigenic specificities; (4) the specificity of the antibodies in a serum is to be compared with the known specificity of the antibodies of a "control" serum (Janeway and Travers, 1996).

Many autoantibodies occur in the sera of patients with autoimmune connective tissue diseases. Most of them recognize nuclear or cytoplasmatic constituents which are ribonucleoproteins; these are complexes of small nuclear or cytoplasmatic RNAs with several proteins. Four autoantigens, the Ro (SSA), La (SSB), Sm and U1RNPs are of considerable importance for diagnostic and/or prognostic purposes for patients with autoimmune connective tissue diseases, in particular systemic lupus

* Department of Pathophysiology, University of Athens, Medical School, 75 Mikras Asias St., 11527 Athens, Greece; Tel.: (+30)-1-7789480; Fax: (+30)-1-7703876; e-mail: pvlah@atlas.uoa.gr

erythematosus (SLE), mixed connective tissue disease (MCTD), Sjogren's syndrome (SS) and rheumatoid arthritis (RA) (Tan 1989). The antigens occur in many tissues, and the Sm and U1RNPs in particular are highly conserved throughout evolution. In clinical practice there is a need for large amounts of the above antigens, occurring in crude tissue extracts, in order to perform either the Ouchterlony gel diffusion assay or counterimmunoelectrophoresis (CIE) to detect autoantibodies in the sera of patients with the above cited autoimmune diseases (Kurata and Tan 1976). Tissues relatively easily to obtain and rich in the above autoantigens are human spleen, calf thymus or rabbit thymus. For specific purposes, such as the RNA precipitation assay and Western blotting, the source of the antigens could be cytoplasmatic (in the case of Ro/SSA and La/SSB) or nuclear (in the case of Sm or U1RNP) HeLa cell extracts (Verheijen et al. 1993).

6.1
Preparation of Human Spleen Extract

Materials

- Tissue homogenizer (blender)
- Human spleen (approximately 130 g): Immediately after excision, the spleen should be placed on dry ice and maintained this way until received from the laboratory; it should then be kept at $-70\,°C$ until use.
- Phosphate buffered saline (PBS), 0.15 M, pH 7.2: To make up 1 l, dissolve 8.0 g NaCl, 2.0 g KCl, 1.15 g Na_2HPO_4, 0.2 g KH_2PO_4 in 1000 ml of distilled H_2O. Adjust the pH.
- Centrifuge
- Ultracentrifuge
- Dialysis tubing: To prepare the tubing, wet it by putting it in distilled H_2O for 30 min just before use.
- Phosphate buffered saline (PBS) 0.5 M, pH 7.2: To make up 1 l, dissolve 29.2 g NaCl, 0.2 g KCl, 1.15 g Na_2HPO_4, 0.2 g KH_2PO_4 in 1000 ml of distilled H_2O. Adjust pH.
- DE-52 cellulose-dry powder
- XM-50 Diaflo-Amicon concentrator

Procedure

The whole procedure should be carried out at a temperature very close to 0 °C and no higher than 4 °C.

1. Place the spleen in the bath of the tissue homogenizer; add ice cold PBS 0.15 M, pH 7.2 (volume of PBS, in milliliters, three times the weight of the spleen in grams).

2. Homogenize the spleen.

3. Centrifuge the mixture at 10 000 g for 30 min.

4. Ultracentrifuge the supernatant at 35 000 g for 3 h.

5. Place the supernatant in the dialysis tubing and dialyze overnight against PBS 0.15 M, pH 7.2.

6. Dissolve 100 g DE-52 cellulose dry powder in 300 ml PBS 0.15 M, pH 7.2.

7. Stir for 30 min and then allow to settle. Measure the pH of the supernatant.

8. Repeat several times until the pH of the supernatant is 7.2; remove the supernatant.

9. Mix the human spleen extract (HSE) with the DE-52 cellulose by stirring overnight.

10. Centrifuge at 10 000 g for 1 h and remove the supernatant.

11. Mix the precipitate with PBS 0.5 M, pH 7.2. Stir for 4 h.

12. Centrifuge at 10 000 g for 1 h.

13. Dialyze the supernatant overnight against PBS 0.15 M, pH 7.2.

14. Concentrate 10× with the XM-50 Diaflo-Amicon system. The extract contains Ro/SSA, La/SSB, Sm and U1RNP antigens (Maddison et al. 1985).

6.2
Preparation of Calf Thymus Extract

Materials

- Tissue homogenizer (blender)
- Fresh calf thymus (approximately 100 g) from the abattoir either put in a plastic bag and keep on ice until received from the laboratory or keep at $-70\,°C$ until use.
- PBS 0.15 M, pH 7.2 (ice cold); for recipe, see above
- Centrifuge
- Ultracentrifuge
- Dialysis tubing
- Diaflo-Amicon

Procedure

1. Remove the fat from the thymus.

2. Add ice cold PBS 0.15M, pH 7.2 (volume of PBS, in milliliters, three times the mass of the thymus in grams).

3. Homogenize the thymus.

4. Centrifuge at 10 000 g for 1 h.

5. Take the supernatant and ultracentrifuge at 100 000 g for 90 min. Take the supernatant.

6. Concentrate 10× using XM-50 Diaflo-Amicon system.

7. Check the extract for its antigenic activity by using prototype sera: It should contain the antigens Ro/SSA, La/SSB, Sm, U1RNP and possibly other not recognized or well defined antigens.

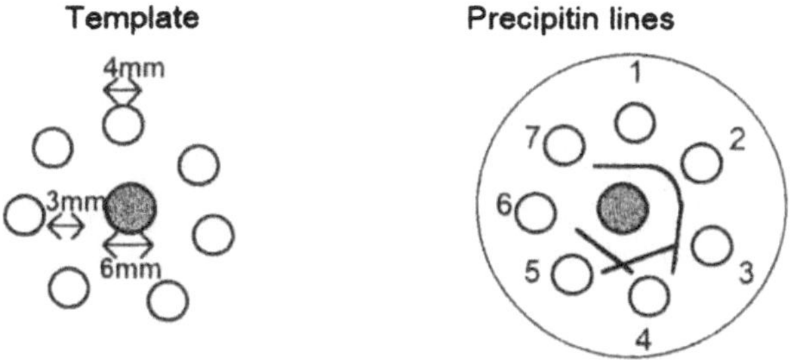

Template   Precipitin lines

Fig. 6.1. Ouchterlony gel immunodiffusion. The serum in well (2) is of unknown specificity. This serum presents a precipitin line which shows continuity with the precipitin lines obtained with known sera in wells (1) and (3). This means that the serum in well (2) has the same antibodies as the sera in wells (1) and (3). By contrast, the sera in wells (4) and (5) have crossing precipitin lines. This means that these sera recognize different antigens in the antigen source (central well); therefore, these sera contain different antibodies. The sera in wells (3) and (4) present precipitin lines which do not cross each other completely, but the precipitin line of serum in well (4) has some continuity with that of serum in well (3). This means that the antibodies in those sera recognize two antigens which have some cross-reactivity or that there are two different antigenic epitopes on the same molecule

6.3
Ouchterlony Immunodiffusion

Materials

- Petri dishes
- PBS 0.15 M, pH 7.2; for recipe, see above
- Hot plate stirrer
- Agarose, high purity
- Pasteur pipette connected to a vacuum line
- Paper template as shown in Fig. 6.1
- Cylindrical metal cutters: one 6 mm and another 4 mm in diameter (to cut wells in the agar plates; Fig. 6.1)
- Humidified chamber
- "Prototype" sera
- Sodium azide

Procedure

1. Prepare 0.6 % agarose in PBS 0.15 M, pH 7.2, containing 0.01 % sodium azide.

2. Pour 8 ml melted agarose into each Petri dish and allow to cool and solidify before cutting the wells. Store the dishes in a humidified chamber at 4 °C and use them within 2 days.

3. Using a paper template, cut a central well 6 mm in diameter, and then on the periphery seven wells 4 mm in diameter (for the sera); these should be 3 mm apart from the central well (Fig. 6.1).

4. Remove the agarose plugs by using a Pasteur pipette connected to a vacuum line

5. Fill the central well with the antigen source (100 μl) and the peripheral wells with the sera under study (20 μl). In the first experiment, we identify the positive sera against one of the antigens included in the antigen source. The positive sera form a precipitin line when they diffuse towards the antigen. In the second experiment, we identify the specificity of the positive sera, by filling, for example, wells 1, 3 and 5 with prototype sera (containing antibodies of known specificities) and the remaining wells, containing the positive sera, with the precipitin lines of unknown specificities from the first experiment (see Fig. 6.1).

6. Place the dishes in a humidified chamber and read the results after 24 and 48 h (Maddison et al. 1985).

6.4
Counterimmunoelectrophoresis

Materials

- Agarose (high purity)
- PBS 0.15 M, pH 7.2; for recipe, see above
- Barbital buffer, ~0.065 M, pH 8.2: To prepare 1 l, dissolve 10.3 g sodium barbital in 800 ml distilled H_2O. Dissolve 1.84 g barbital in 200 ml distilled H_2O after heating. Mix the solutions and adjust the pH
- Glass plates 8.5×8.5 cm, alcohol-cleaned and dry
- Hot plate stirrer
- Paper template, as shown in Fig. 6.2
- Cylindrical metal cutter and a parallelogramic metal cutter to cut wells (4 mm diameter) and troughs (3 mm wide) on the agar plate according to Fig. 6.2

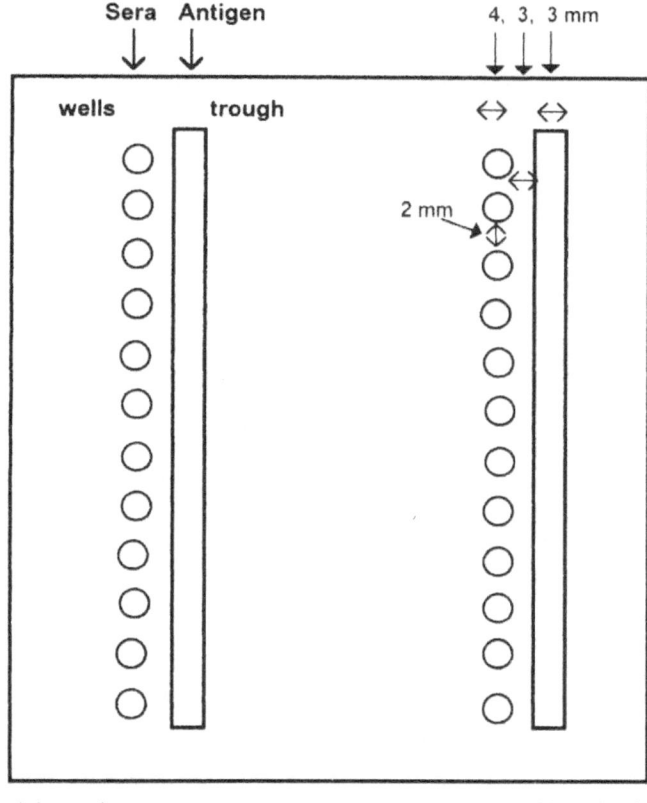

Fig. 6.2. A typical template for a counterimmunoelectrophoresis (CIE) agarose electrophoresis plate. Sera are placed in the wells and antigen in the troughs. The precipitin lines are formed in the 3 mm space between the trough and the row of wells. Alternatives to this experiment include the following: (1) the diameter of the wells and troughs can be decreased and a third row added in the middle; (2) instead of troughs, wells for the antigen, in a parallel row to the ones for the sera, can be used. However, using troughs it is easier to obtain identity lines between adjacent wells of sera in order to study antigenic specificities of the antibodies in the sera. **Note:** Marking the glass plates with a permanent label (i.e., upper left corner scrach initials, numbers or dates with a diamond knife) is recommended in order to remember the orientation and to have permanent records of the plates

- Pasteur pipette connected to a vacuum line
- Electrophoresis apparatus and power supply
- Coumassie blue stain: Dissolve 1 g of Coumassie brilliant blue-R-250 in 100 ml glacial acetic acid, 450 ml ethanol and 450 ml distilled H_2O.
- Destain solution: Mix 100 ml glacial acetic acid, 250 ml ethanol and 650 ml distilled H_2O.

Procedure

1. Prepare 1 % agarose by dissolving 1 g agarose in 100 ml of barbital buffer, ~0.065 M, pH 8.2, by heating on a hot plate stirrer.

2. Put the glass plates on a horizontal leveling table and pour 12 ml agarose per plate. Leave the agarose to cool. Use the plates 2 h after preparation or within 2 days, provided that the plates have been kept in a humidified chamber at 4 °C.

3. Cut 4 mm diameter wells, arranged in two parallel rows, over the length of the plate. Each row should contain 12 wells. Neighboring wells in the same row are separated by a distance of approximately 2 mm. Parallel to each row, to the cathode side, cut a 3 mm wide trough, separated from each row of wells by 6 mm. Remove the agarose plugs from the wells but not from the troughs.

4. Load 20 µl serum into each well, transfer the plate to the electrophoresis apparatus containing barbital buffer, ~0.065 M, pH 8.2. Electrophorese the samples at constant current, 12 mA per plate, for 15 min. The wells with the sera are placed towards the anode.

5. After 15 min, turn off the current, remove the gel plugs from the troughs and fill them with antigen (250 µl of freshly thawed antigen preparation).

6. Continue electrophoresis for 60 min using the same current and polarity.

7. Read the results.

8. Place the plate in a humidified chamber at 4 °C and read the results after 24 h.

9. Wash the plate in cold PBS 0.15 M, pH 7.2, for 8 h (overnight).

10. Stain the plate with Coomassie blue in order to have permanent records of the precipitation lines. Remove the background color with the destain solution and the precipitation lines will remain stained permanently in blue.

6.5
Preparation of an Immunoaffinity Column for Purification of Small Nuclear Ribonucleoprotein Particles (snRNP Peptides)

Materials

- Purified IgG (~90 mg) from a serum with anti-U1RNP antibodies, negative for other autoantibodies. IgG can be purified according to standard protocols using a protein-A Sepharose column
- Cyanogen bromide (CNBr)-activated Sepharose-4B
- Carbonate-bicarbonate 0.1 M/glycine 0.2 M buffer, pH 8.3: To prepare this buffer, make stock solutions of 1 M sodium carbonate and 1 M sodium bicarbonate. Mix carbonate and bicarbonate solutions in the proportions 1:9 (v/v) and adjust to pH 8.3 by pouring an appropriate amount of one of the above solutions into the buffer. Add glycine to the final volume of the buffer in order to reach a molarity of glycine equal to 0.2M. Adjust the pH to 8.3. **Note:** Carbonate solution raises the pH, bicarbonate solution lowers the pH
- Acetate 0.1 M in 0.5 M NaCl, pH 4.0
- Acid solution, HCl 1 mM
- PBS 0.15 M, pH 7.2; for recipe, see above
- Column: 3.5 cm internal diameter and 20 cm long
- Tubing: 2 mm internal diameter
- Container of approximately 800 ml volume with a glass funnel shape and a porous bottom filter
- Plastic tube, with at least a volume of 50 ml, with stopper
- End-over-end stirrer
- Ultraviolet (UV) detector analyzer
- Chart recorder
- Cold room or cold cabinet
- Centrifuge

- Peristaltic pump in order to provide the flow rate of the crude antigen extract through the column. Alternatively, you can use the hypsometrical difference between the container with the crude antigen extract (higher off the ground) and the container collecting the eluted antigen through the column (lower than the crude antigen extract container)
- Vacuum line

Procedure

1. Put 4 g CNBr-activated Sepharose-4B (dry powder) in a container (glass funnel shape) with a porous bottom. Swell the gel with 1 mM HCl (200 ml/g of gel) under continuous stirring for 15 min.

2. Connect the funnel shaped container with a vacuum line to remove the liquid phase of the suspension through the porous bottom of the container.

3. Repeat gel washing four times with 1 mM HCl.

4. Collect the swollen gel layer from the bottom of the container and put the gel in a tube. You will have 15 ml of gel.

5. Mix the preparation of purified IgG with the gel (5 mg IgG/ml gel) by filling the tube with the IgG solution. Close the tube using the stopper and mix by using an end-over-end stirrer, very gently, at room temperature for 2 h.

6. Put the suspension in the same container as before and remove fluid and gas using the vacuum line. Check the fluid for the presence of IgG by immunonephelometry. If the coupling of IgG to CNBr-activated Sepharose-4B is complete, you will have only a trace of IgG (less than 0.5 % of the initial amount).

7. Remove the gel from the container. Put the gel in the tube with 30 ml carbonate-bicarbonate 0.1 M/glycine 0.2 M buffer, pH 8.3, and mix at room temperature using the end-over-end stirrer for 2 h.

8. Put the suspension in the container and remove the fluid using the vacuum line. Mix the gel in the container, first with 0.1 M $NaHCO_3$ in 0.5 M NaCl, pH 8.3, and then, after removing the liquid phase of the suspension, with 0.1 M acetate buffer in 0.5 M NaCl, pH 4.0.

9. Repeat four times using the same interchange of buffers. Remove the fluid phase of the suspension.

10. Dissolve the gel in PBS 0.15 M, pH 7.2, with 0.02 % sodium azide (to stop microbial growth), and remove the fluid phase of the suspension; repeat three times.

11. Dissolve the gel in PBS 0.15 M, pH 7.2, with 0.02 % sodium azide and gently pour the gel into the column slowly and continuously. Leave the column outlet open during packing. The buffer will pass through the column and the Sepharose, coupled with anti-U1RNP IgG, will settle. Collect the eluent from the column and check for IgG using a nephelometer. IgG should be absent in the eluent.

12. Wash the column with PBS 0.15 M in 0.5 M NaCl, pH 7.2 (13 mM Na_2HPO_4, 3 mM KH_2PO_4, 0.5 M NaCl). Use 300–500 ml of this buffer.

13. Wash the column with PBS 0.15 M, pH 7.2, with 0.02 % sodium azide.

14. Keep the column equilibrated with PBS 0.15 M, pH 7.2, with 0.02 % sodium azide at 4 °C until use. Block the outlet of the column to prevent the material from drying. Cover the top of the column with the appropriate cap. Ensure that a flow adaptor is incorporated into the cap to connect the chamber containing the crude antigen preparation with the column. Ensure that a small amount of buffer remains on top of the gel in the column.

6.6
Purification of SnRNP Peptides from Crude Tissue Extracts by Affinity Chromatography

Materials

- Cold room or cold cabinet
- Calf thymus extract (CTE) enriched in snRNP peptides
- Affinity column prepared as previously described
- UV analyzer
- Chart recorder

- Peristaltic pump (or use the hypsometrical difference between the chamber containing the antigen extract and the chamber collecting the material coming through the column)
- Dialysis tubing
- PBS 0.15 M, pH 7.2
- PBS 0.5 M, pH 7.2
- Guanidine-HCl 3.0 M, pH 7.2, dialyzed in PBS: Into 1 l of PBS 0.15 M, pH 7.2, add an amount of guanidine-HCl in order to prepare a 3.0 M solution.
- Ammonium sulfate

Procedure

Enrichment of CTE in snRNPs by Ammonium Sulfate Fractionation

The procedure should take place at 4 °C.

1. Take the crude CTE before final concentration(see steps 1–3, Sect. 6.2). Add 16.4 g ammonium sulfate/100 ml CTE. Mix for 30 min. Spin at 5000 g for 30 min; save the supernatant at 4 °C.

2. Dilute the precipitate in 30 ml PBS 0.15 M, pH 7.2, and pour the solution into dialysis tubing 2 (tubing volume should be three times larger than the volume of the solution). Dialyze overnight against 6 l PBS 0.15 M, pH 7.2. The final solution is the 0–30 % ammonium sulfate CTE fraction.

3. To the supernatant of step 1, add 18.1 g ammonium sulfate/100 ml. Stir for 30 min. Centrifuge at 5000 g for 30 min and keep the supernatant. Take the precipitate and dilute it in 30 ml PBS 0.15 M, pH 7.2. Dialyze overnight against 6 liters PBS 0.15 M, pH 7.2. This is the 30 %–60 % ammonium sulfate fraction.

4. Check for the presence of Sm and U1RNP antigens in this fraction by using CIE or Ouchterlony immunodiffusion and prototype anti-Sm/anti U1RNP sera.

Affinity Purification of Sm/U1RNP Complex

1. Pass the 30 %–60 % ammonium sulfate CTE fraction through the anti-U1RNP affinity column, with a rate of 15 ml/ha.

A first peak of unbound proteins is shown through the chart recorder. When the peak returns to the baseline, wash out the remaining unbound proteins with 100 ml PBS 0.5 M, pH 7.2, and subsequently with 200 ml of PBS 0.15 M, pH 7.2.

2. When the chart recorder indicates baseline, wash the column with 3.0 M guanidine-HCl, pH 7.2, and collect the fraction shown by the chart recorder. This fraction will contain the Sm/U1RNP antigen. Dialyze this fraction overnight at 4 °C against 6 l ice cold PBS 0.15 M, pH 7.2.

3. Check the antigenicity by CIE. Keep in small aliquots (10–20 µl) at −70 °C until use (Maddison et al. 1985).

6.7
ELISA for Antibodies to U1RNP

Materials

- Ninety-six well polystyrene ELISA plates
- Poly-L-lysine
- Affinity purified antigen
- PBS 0.15 M, pH 7.2
- Bovine serum (BS)
- Tween 20
- p-Nitrophenyl phosphate disodium salt
- Alkaline phosphatase-conjugated anti-human γ, α, or µ chain-specific antisera
- Diethanolamine buffer, pH 9.8
- Incubator
- Parafilm
- Control sera anti-U1RNP-positive and 12 normal sera per plate

Procedure

1. Put in the wells 100 µl of poly-lysine solution (50 µg/ml in PBS). Incubate at 37 °C for 30 min after covering the plates with parafilm. Discard any solution not bound to the plate.

2. Apply purified antigen (100 µl of a 20 µg/ml solution in BS 10 % in PBS 0.15 M, pH 7.2) and incubate at 37 °C for 30 min. Discard any solution not bound to the plate.

3. Put 100 μl of BS 10 % in each well and incubate for 1 h at 37 °C to block the nonspecific binding sites.

4. Into each well, place 100 μl of serum, diluted 1:1000 with BS 10 % in PBS 0.15 M, pH 7.2, to detect IgG, and 1:100 for the detection of IgA and IgM antibodies; incubate for 2 h.

5. Wash the plates with PBS-Tween 20 (0.05 % Tween in PBS 0.15 M, pH 7.2).

6. Into each well, place 100 ml/well alkaline phosphatase-conjugated anti-human g, a, or m chain-specific antisera, diluted 1:1000 in BS 10 %; incubate for 1 h at 37 °C.

7. After extensive washing with PBS-Tween 20, add *p*-nitrophenyl substrate solution in diethanolamine buffer, pH 9.8 (1 mg/ml); stop the reaction after 30 min. Read the results in an ELISA reader at 405 nm. Positive results are those with absorbance values higher than the mean of normals increased by 3 standard deviations ($>x + 3$ SD).

References

Janeway CA, Jr. and Travers P. (1996) Immunobiology, Current Biology Ltd. UK, pp. 2.13–2.14

Kurata N, Tan EM (1976) Identification of antibodies to nuclear acidic antigens by counterimmunoelectrophoresis. Arthr Rheum 19:574–579

Maddison PJ, Skinner RP, Vlachoyiannopoulos P, Brennand D, Hough D (1985) Antibodies to nRNP, Sm, Ro(SSA) and La(SSB) detected by ELISA: their specificity and inter-relations in connective tissue disease sera. Clin Exp Immunol 62:337–345

Ouchterlony O and Nilsson L-A (1978) Immunodiffusion and immunoelectrophoresis. In: Weir DM (ed) Handbook of experimental immunology, vol 1, 3rd edn. Blackwell Scientific, Oxford

Tan EM (1989) Antinuclear antibodies: diagnostic markers for autoimmune disease and probes for cell biology. Adv Immunol 44:93–151

Verheijen R, Salden M Van Venrooij WJ (1993) Protein blotting. In: Manual of biological markers of disease, vol 14. Kluwer Academic, Dordrecht, pp 1–2

Chapter 7

Methods in Immunolocalization of Autoantigens

MARTIN BLÜTHNER*

Introduction

What Are Autoantibodies?

The immune system recognizes and eliminates foreign antigens by a complex network of systems. To effectively fulfill this task the immune system has to discriminate between foreign and self antigens. The mechanisms by which this discrimination is accomplished shall not be discussed here (there are numerous excellent textbooks on immunology available). However, in auto-immune disorders this discrimination fails to a certain extent due to mechanisms currently only vaguely understood. Linked with these diseases is the occurrence of circulating autoanti-bodies. These are antibodies directed against self antigens of a defined nature. Autoantibodies may have a direct effect on the etiology of the disease, such as in Grave's disease. Here, the tar-get antigen of the autoimmune response is the acetylcholine receptor of the motoric endplate; this leads to a severe distur-bance in the nervous system. In other cases, the so-called sys-temic autoimmune disorders, the connection between etiology and target antigen is not obvious. To what extent these autoanti-bodies are involved in the pathogenesis of the corresponding disease is not exactly known. In the systemic diseases the target antigen commonly resides within the cell, especially within the cell nucleus. Several of these anti-nuclear antibodies (ANAs) are significantly linked to specific diseases. Hence, they are ascribed a marker function for these specific diseases. Systemic lupus ery-thematosus (SLE), e.g., is accompanied by the occurrence of

* Institute of Molecular Genetics, University of Heidelberg,
  Im Neuenheimer Feld 230, 69120 Heidelberg, Germany;
  Tel.: (+49)-6221–54–5649; Fax: (+49)-6221–54–5678;
  e-mail: bluthner@sirius.mgen.uni-heidelberg.de

Table 7.1. Features of some of the most important nuclear autoantigens[a]

Antigen	Molecular identity	Disease	Indirect cytoimmunofluorescence	Western blot
Sm	Core proteins of snRNP's	SLE	Speckled nuclear staining, no staining of nucleoli	B-protein (26–20 kD[a])
RNP	U1-snRNP	SLE overlap	Similar to Sm, occasionally more granular	70 kD[a] 30 kD[a] proteins
Scl-70	Topoisomerase I	Scleroderma	Fine speckled nuclear staining, frequent staining of the nucleolus	100 kD[a] protein (degradation product of 70 kD[a])
CENP	Centromere-associated proteins CENP-A, B, C	Scleroderma	Centromere staining	CENP-A (19.5 kD[a]) CENP-B (80 kD[a])
Fibrillarin	Protein component of U3-snoRNP	Scleroderma	Clumpy, nucleolar staining	36 kD[a] protein
PM/Scl	Nucleolar particle of unknown function	Polymyositis scleroderma overlap	Homogeneous, nucleolar staining	100 kD[a], 75 kD[a] proteins some smaller proteins
Ro (SSA)	Protein component of hY-RNP	Sjögren's syndrome	Fine, granular nuclear staining	60 kD[a], 52 kD[a] proteins
La (SSB)	RNA-polymerase III associated factor	Sjögren's syndrome	Fine, granular nuclear staining	48 kD[a] protein
AMA	Subunits of mitochondrial 2-oxo-acid dehydrogenase complex	PBC	Large, granular cytoplasmic staining frequently accompanied by multiple nuclear speckles (sp100 antigen)	70–74 kD[a] protein frequently 100 kD[a] (sp100 antigen)

[a] This is not a comprehensive list; only the most prominent features are listed.

autoantibodies directed against an antigen called the Sm antigen. The antigen consists of several core proteins of the small nuclear ribonucleoprotein particles (snRNPs). Other marker autoantibodies include anti-topoisomerase I (progressive systemic sclerosis), anti-mitochondrial antibodies (primary biliary cirrhosis), or anti-fibrillarin (diffuse scleroderma). Very often the respective autoantigen is only known by its cellular and/or biochemical data but not by its biological function, such as the PM/Scl antigen. A review on the most important autoantigens in various autoimmune diseases is given by Tan (1989). In addition, a summary of the cellular and molecular data of these autoantigens can be found in a book edited by van Venrooij and Maini (1994). A short description of the properties of important autoantigens is given in Table 7.1.

Some autoimmune disorders are further grouped into several subclasses which are defined by a different spectrum of marker antigens recognized by their corresponding autoantibodies. Thus, it becomes evident that elucidation of the specific autoimmune repertoire is crucial for the correct diagnosis of a certain autoimmune disorder. Additionally, for practical purposes, it is faster, cheaper and less laborious to determine the specificity of the autoimmune response than to determine several clinical parameters.

Yet, since some autoantibodies display similiar immunoreactions in diagnosis, it is important to have available several independent test systems which are additionally very sensitive, specific and unequivocal in their detection of autoantibodies. In comparing the results of these independent tests one can establish a clear diagnosis of the underlying autoimmune disorder.

In addition to their clinical relevance and in view of their unique specificities autoantibodies also can serve as valuable tools in basic research. They have been successfully used in dissecting structural and functional features of subcellular components such as snRNPs.

Principle of Immunodetection of Autoantibodies

In the following section we will present some of the most important and most common techniques to detect autoantibodies. The techniques provided include indirect cytoimmunofluorescence, Western blotting and enzyme-linked immunosorbent assay (ELISA). In addition, some special applications such as prepara-

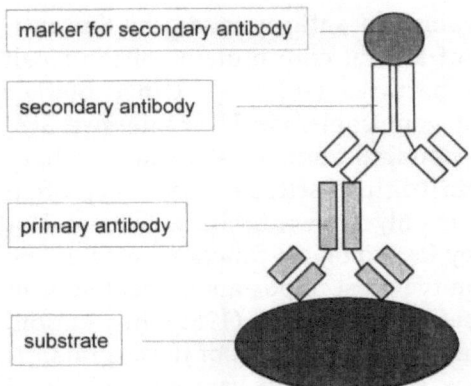

marker for secondary antibody

secondary antibody

primary antibody

substrate

Fig. 7.1. Principle of immunodetection of autoantigens. A substrate is allowed to react with it's corresponding antibody. Bound antibodies are detected with a secondary antibody usually directed against the Fc portion of the primary antibody. The secondary antibody is coupled to a marker. This marker may be either a fluorescent dye, as in indirect cytoimmunofluorescence, or, as in Western blotting or ELISA, an enzyme. The enzyme, usually either horseradish peroxidase or alkaline phosphatase, when incubated with its appropriate substrate, gives a color reaction, thereby indicating the position of the antigen

tion of subcellular protein fractions and affinity purification of autoantibodies are presented. Certainly, these techniques are also applicable to the detection of other antigen/antibody systems, for example in studying infectious diseases, or in basic research as well.

The very principle of most immunodetection procedures is depicted in Fig. 7.1. A substrate is prepared depending on the question asked in the specific experiment. The substrate containing the antigen is allowed to react with a primary, specific antibody. This primary antibody can be either a serum, an affinity-purified antibody or a monoclonal antibody. After washing off of nonspecifically bound antibodies the remaining specific antibodies, which are now bound to their antigen, are detected with a secondary antibody, usually directed against the Fc portion of the primary antibody. The secondary antibody is coupled to a detection reagent such as a fluorescent dye for direct visualization or an enzyme to perform a color reaction with its corresponding substrate. Most of these secondary antibodies are available commercially in already labeled form. Depending on the type of immunoassay used, the results have to be assessed visually (e.g., indirect cytoimmunofluorescence or Western blot) or can be quantitated directly (e.g., ELISA).

7.1
Indirect Cytoimmunofluorescence

Autoantibodies often recognize distinct subcellular structures such as snRNPs, also known as Sm antigen or RNP antigen (see Sect. 7.2), topoisomerase I, also known as Scl-70 antigen, or fibrillarin, a nucleolar protein associated with the U3 snoRNP. For clinical analysis it is critical to determine the specificity of the antibody in order to define the underlying autoimmune disease. Therefore, a first step in the analysis of a given autoimmune disease is indirect cytoimmunofluorescence to localize the target antigen within the cell. The antibody is allowed to react with a cellular substrate and thereafter is visualized with a secondary anti-antibody coupled to a fluorescent dye, thus allowing localization of the subcellular compartement harboring the antigen against which the autoantibody is directed (Fig. 7.2). A crucial step in indirect cytoimmunofluorescence is fixation and permeabilization of the cellular substrate. Numerous methods are used; however, in our hands, the most reliable method is fixation with methanol and permeabilization with acetone, resulting in the lowest background. An experienced clinician can readily determine the specificity of an autoantibody by its staining pattern in

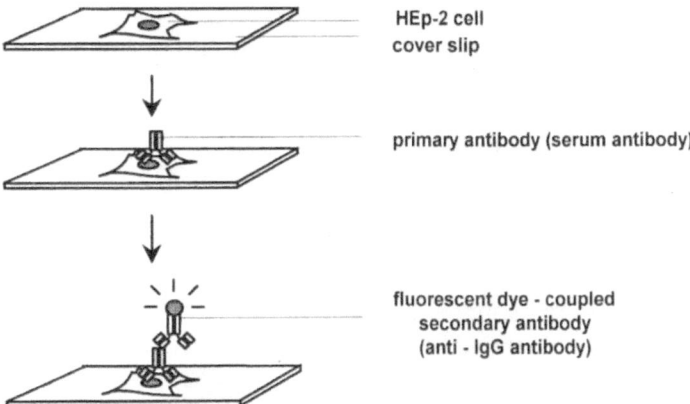

HEp-2 cell
cover slip

primary antibody (serum antibody)

fluorescent dye - coupled
secondary antibody
(anti - IgG antibody)

Fig. 7.2. Principle of indirect cytoimmunofluorescence. HEp-2 cells are grown on cover slips. After fixation and permeabilization serum antibodies are incubated with the cells. Bound antibodies are detected with a secondary antibody directed against the Fc portion of the primary antibody, which is coupled to a fluorescent dy such as FITC or rhodamine. Under a microscope equipped with the appropriate filters the position of the antigen thus lights up

indirect cytoimmunofluorescence. Anti-Sm antibodies, e.g., are characterized by giving a fine speckled nuclear staining pattern, as opposed to a more granular staining pattern caused by anti-U1 RNP antibodies. Anti-fibrillarin antibodies stain the nucleolus in a clumpy pattern as opposed to anti PM/Scl antibodies, which give a homogeneous nucleolar staining. These few examples already provide an idea of the importance of determining the subcellular location of a given autoantigen.

As a common substrate for indirect cytoimmunofluorescence the permanent cell line HEp-2 is used. The cell line is derived from a human larynx carcinoma. The cells are easy to handle and to maintain. They grow in a monolayer and exhibit an epithelial morphology. These features are particularly useful for cytoimmunofluorescence since the cells are flattened out, making it easy to focus with the microscope on a given staining pattern. An additional advantage is the presence of several nucleoli, making the cells particularly useful for the detection of nucleolar autoantigens.

Materials

- HEp-2 cells (ATCC# CCL 23)
- Cover slips
- Glass slides
- Microscope equipped with filters for rhodamine- and FITC detection
- Plastic tray
- Whatman 3MM filter paper
- Tapered forceps
- Secondary antibody (e.g., FITC- or rhodamine-coupled goat anti-human IgG)

Buffers and solutions
- Phosphate buffered saline(PBS): 8 mM Na_2HPO_4; 1.5 mM KH_2PO_4; 140 mM NaCl; 2.6 mM KCl. Adjust pH to 7.3
- Methanol (prechilled to $-20\,°C$)
- Acetone (prechilled to $-20\,°C$)

Procedure

1. Sterilize a few cover slips by dipping them with forceps in ethanol and flaming them over a Bunsen burner.

2. Use forceps to place cover slips in a plastic dish.

3. Seed HEp-2 cells at a density of 2×10^4/ml in the plastic dish. We grow HEp-2 cells in RPMI medium and seed them in a volume of 50 ml in plastic dishes with a diameter of 13 cm. Thus, we seed routinely 1×10^6 cells.

4. Grow cells for approximately 24 h at 37 °C. By then cells are grown to an optimal density for indirect cytoimmunofluorescence.

5. Using tapered forceps dip the cover slip briefly in PBS to wash off any adhering culture medium.

6. Fix the cells by incubation in -20 °C cold methanol for 5 min.

7. Dip the cells for 5 s in -20 °C cold acetone. This step permeabilizes the cytomembrane to make the intracellular matrix accessible to antibodies.

8. Air-dry briefly

9. Prepare a humidified chamber. This is done by placing a moistened (not wet) filter paper in a plastic dish.

10. Place the cover slip face-up on the moistened filter paper.

11. Add 30 µl of your diluted antibody in PBS to the cover slip. A good dilution to start with when using autoimmune sera is 1:100. The dilution factor certainly depends on the titer and the affinity of a given antibody and has to be determined individually for each serum.

12. Place the lid back on the plastic tray and incubate for 30 min at room temperature.

13. Dip the cover slip briefly in PBS to remove excessive antibody solution and thereafter wash three times for 10 min in PBS.

14. Return cover slip to the humidified chamber and carefully overlay with 30 µl of the FITC- or rhodamine-coupled secondary antibody dilution (1:100 in PBS).

15. Place the lid back on the plastic tray and incubate for 30 min at room temperature. This time the incubation has to be done in the dark to avoid bleaching of the light-sensitive fluorescent dye. We simply cover the tray with a cardboard box.

16. Remove excessive antibody and wash as described in step 13, still keeping the cover slips in the dark.

17. To mount the cover slips, place a drip of mounting medium (e.g., Mowiol or VectaShield) on a glass slide and carefully place the cover slip, this time face down, on the glass slide. The slides are now ready to be examined under the microscope with the appropriate fluorescence filter.

7.2
Preparation of Cytoplasmic, Nuclear and Nucleolar Fractions for Western Blot

After localization in cytoimmunofluorescence it is often necessary to further characterize the molecular features of an autoantigen. This is preferably done by Western blotting (see Sect. 7.4). However, few autoantigens are readily detectable in Western blot experiments using whole cell extracts. Since the antigens are often low abundant proteins or particles, it is usually neccessary to enrich and partially purify the antigens. The following section describes a procedure used in our lab to prepare a cytoplasmic fraction, different nuclear fractions and a nucleolar fraction (Fig. 7.3; Guldner et al. 1983; Blüthner and Bautz 1992). For preparation of the subcellular fractions we use HeLa S3 cells. The cell line is a derivative of the classical HeLa line (Gey et al. 1952), derived from a human cervical carcinoma, but is adapted to growth in suspension culture. This feature allows one to grow the cells in the quantities neccessary to prepare the subcellular fractions. The cytoplasmic fraction is used to detect, e.g., mitochondrial antigens such as occur in primary biliary cirrhosis (PBC). The SLE-specific Sm and U1 RNP antigens are detectable in the RNP fraction, whereas the scleroderma-specific anticentromere antibodies react with proteins enriched in the fraction of soluble nuclear proteins. Typical nucleolar autoantigens are fibrillarin, a component of the U3-snoRNP, specific for certain forms of scleroderma, and the PM/Scl-antigen, a nucleolar autoantigen of unknown function, being the target antigen in sera from patients suffering from polymyositis/scleroderma overlap syndrome. This is certainly not a comprehensive list of detectable autoantigens in the subcellular fractions described; the specificity of a given autoimmune serum has to be determined individually.

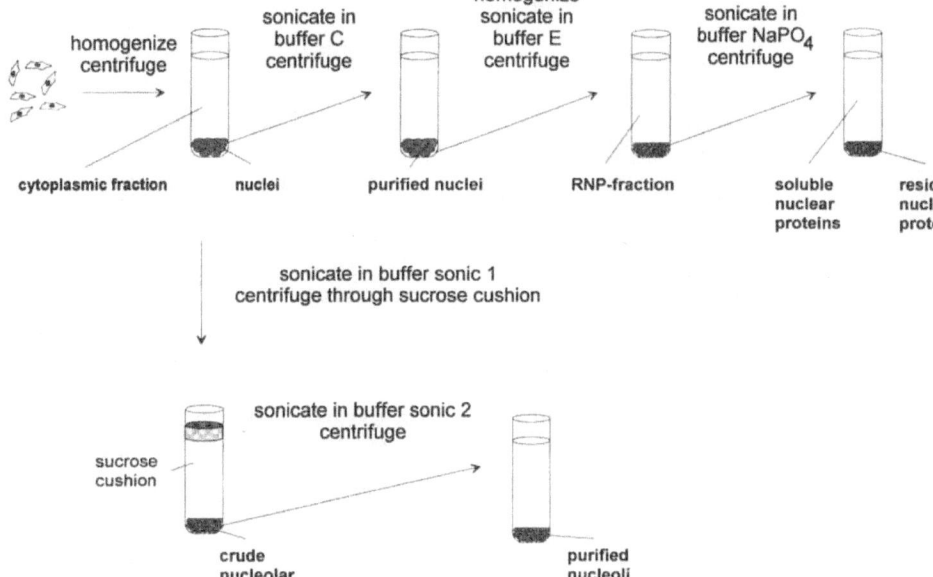

Fig. 7.3. Flow chart for preparation of cytoplasmic, nuclear and nucleolar fractions. HeLa S3 cells are disrupted by homogenization. The supernatant after centrifugation is the cytoplasmic fraction, the pelleted nuclei are further processed to prepare nucleoli or various nuclear fractions. For preparation of nucleoli the nuclei are sonicated and centrifuged through a sucrose cushion followed by an additional sonication and centifugation step. The nuclear fractions are obtained by a series of sonication and centrifugation steps in different buffer systems

▒ Materials

– Glass homogenizers L (loose fitting), S (tight fitting) (Braun, Melsungen, Germany)
– Corex glass tubes 15 ml and 30 ml
– Sonifier (model B 12) equipped with a microtip (Branson)
– Microscope (model III, Zeiss, Oberkochen, Germany) equipped with an UV-filter
– Centrifuge equipped with fixed angle and swing-out rotor, e.g., Sorvall or Beckman
– HeLa S3 (ATCC# CCL 2.2): a HeLa strain adapted to growth in suspension culture

Buffers and solutions

Preparation of nuclei and cytoplasmic extract:

- Phosphate-buffered saline (PBS): 8 mM Na_2HPO_4; 1.5 mM KH_2PO_4; 140 mM NaCl; 2.6 mM KCl. Adjust pH to 7.3.
- Buffer A: 10 mM Tris-HCl, pH 7.0; 10 mM NaCl; 1.5 mM $MgCl_2$; 0.1 % Triton X-100
- Buffer B: same as buffer A but without Triton X-100
- TCA: 10 % stock solution in water
- Acetone (prechilled to −20 °C)

Preparation of subnuclear fractions:

- Buffer C: 50 mM Tris-HCl, pH 7.5; 25 mM KCl; 10 mM $MgCl_2$; 50 mM EDTA
- Buffer E: 10 mM Tris-HCl, pH 8.0; 140 mM NaCl; 1 mM $MgCl_2$
- $NaPO_4$ buffer: 1 mM Na_2HPO_4; 1 mM NaH_2PO_4. Use 1 mM NaH_2PO_4 solution to adjust 1 mM Na_2HPO_4 to pH 7.0

Preparation of nucleolar proteins:

- Buffer sonic 1: 250 mM sucrose; 10 mM Tris-HCl, pH 7.4; 1 mM $MgCl_2$
- Sucrose cushion: 880 mM sucrose; 10 mM Tris-HCl, pH 7.4
- Buffer sonic 2: 880 mM sucrose; 10 mM Tris-HCl, pH 7.4, 1 mM $MgCl_2$

Note: To all buffers the protease inhibitor PMSF (phenylmethyl sulfonylfluoride) is added just prior to use (final concentration of 0.1 mM).

Procedure

Preparation of Nuclei and Cytoplasmic Extract

1. Harvest approximatly 1.5×10^9 HeLa S3 cells grown in suspension culture by centrifugation at 1000 rpm.

2. Wash cells twice by resuspension in the original volume of cold PBS and centrifugation at 1000 rpm for 10 min.

3. Resuspend the pelleted cells in 30 ml cold buffer A and place the cell suspension in a tight fitting, prechilled glass homogenizer.

4. Place homogenizer on ice and homogenize until virtually all nuclei are microscopically free from cytoplasmic remnants.

This means: do a series of ten strokes, place a drop of the suspension on a glass slide and check nuclei under the microscope. Repeat this procedure two to four times. By then most of the nuclei should be virtually free of cytoplasmic contaminants. Clean nuclei display a smooth surface under the microscope.

5. Collect nuclei by centrifugation at 2000 rpm for 10 min. Set aside the supernatant. The supernatant is the cytoplasmic fraction (see step 7), the pelleted nuclei are washed (see step 6) and further processed for the three different nuclear fractions or the nucleolar fraction (see flow chart).

6. Wash nuclei from step 5 twice in 30 ml buffer B by resuspension and centrifugation at 2000 rpm, 10 min. Proceed to "Preparation of Subnuclear Fractions" or "Preparation of Nucleolar Proteins".

7. Add an equal volume of a 10 % TCA solution to the supernatant and place on ice for at least 1 h.

8. Centrifuge at 10 000 g for 30 min.

9. Wash pelleted material two or three times in a large volume of acetone (prechilled to −20 °C).

10. Air-dry the precipitated protein and resuspend in an appropriate amount of PBS (approximately 2–3 ml). If you need the proteins for Western blotting you can also resuspend them directly in Laemmli sample buffer.

Preparation of Subnuclear Fractions

1. Resuspend the washed nuclei (Preparation of Nuclei and Cytoplasmic Extract, step 6) from above in 20 ml buffer C and place the suspension in a loose fitting, pre-chilled glass homogenizer.

2. Homogenize with 15 strokes.

3. Harvest nuclei by centrifugation at 10 000 g, 10 min. Nuclei are now free of any adhering ribosomes.

4. Resuspend purified nuclei in 2 ml buffer E and keep the suspension on ice for 45 min.

5. The nuclear suspension is sonicated twice for 10 s at 45 W. Between the two sonication steps allow the suspension to cool for at least 30 s.

6. Centrifuge at 10 000 g for 10 min. The resulting supernatant is the enriched RNP fraction.

7. The pellet is resuspended in 10 ml NaPO$_4$ buffer and sonicated ten times 10 s at 45 W. Again, allow the suspension to cool for 30 s between the 10 s sonication steps.

8. Centrifuge the sonicated fraction at 10 000 g for 10 min.

9. The supernatant contains what we call the soluble nuclear proteins. The pellet represents the residual proteins. These residual proteins are resuspended directly in 2 ml Laemmli sample buffer.

To determine the protein yield measure the optical density (OD) of the cytoplasmic fraction, the RNP fraction and the soluble protein fraction against their respective buffers at 260 and 280 nm. The protein concentration can be calculated by the following formula (Kalckar 1947):

$$\text{mg protein/ml} = (1.45 \times OD_{280\,nm}) - (0.74 \times OD_{260\,nm})$$

Keep in mind that the determination of protein concentrations according to this formula is only applicable when the protein solution is a homogeneous mixture of different proteins and should not be applied to purified proteins. Yields between 5 and 15 mg/ml are a reasonable result.

Preparation of Nucleolar Proteins

1. Resuspend the pelleted nuclei (Preparation of Nuclei and Cytoplasmic Extract, step 6) in 15 ml buffer sonic 1.

2. Sonicate five times for 5 s at 45 W with 30 s cooling breaks between the single steps.

3. Prepare three centrifuge tubes with 10 ml sucrose cushion per tube. Overlay the sucrose cushion with 5 ml of the sonicated solution.

4. Centrifuge for 30 min at 2000 g in a swing out rotor.

5. Resuspend the pelleted crude nucleolar fraction in 2 ml buffer sonic 2.

6. Sonicate four times for 5 s as before.

7. Centrifuge for 20 min at 5000 g.

8. Resuspend the pelleted nucleoli directly in 2 ml Laemmli sample buffer.

7.3
Discontinuous SDS Polyacrylamide Gel Electrophoresis

Several procedures for investigating the molecular features of a given antigen depend on the efficient separation of proteins

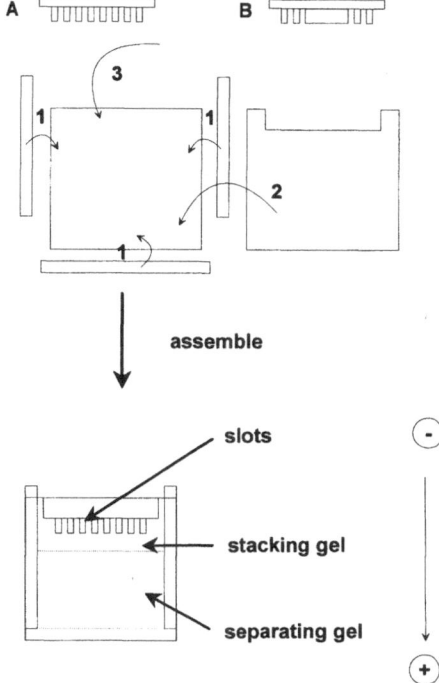

Fig. 7.4. Preparation of a SDS-PAGE gel. Place spacers (*1*) between the two glass plates (*2*) and fix the sandwich with clamps. Seal with 1.5 % agarose. Mix solutions for the separating gel according to Table 7.2. Pour separating gel up to two thirds below the upper rim of the sandwich. Allow to polymerize for at least 1 h. Prepare solutions for the stacking gel according to Table 7.3. Pour the stacking gel, put in the comb (*3*). This may be either a comb forming single slots (*A*) or a comb forming one large, preparative slot with adjacent single slots for molecular weight markers (*B*). Allow to polymerize for at least 1 h. After loading the samples run the gel with 75–100 V until the samples enter the separating gel. Then raise the voltage to 180–200 V. When the bromophenol band reaches the bottom of the gel the run is finished

Table 7.2. Components and volumes of solutions for preparation of SDS-polyacrylamide separating gels

| Component | Acrylamide concentration | | | | | | |
	5 %	7.5 %	10 %	12.5 %	15 %	17.5 %	20 %
30% Acrylamide	5.0 ml	7.5 ml	10.0 ml	12.5 ml	15.0 ml	17.5 ml	20.0 ml
1% Bis-acrylamide	7.8 ml	5.8 ml	3.9 ml	3.1 ml	2.6 ml	2.2 ml	2.0 ml
1.5M Tris/Cl pH 8.7	7.5 ml	7.5 ml	7.5 ml	7.5 ml	7.5 ml	7.5 ml	7.5 ml
H_2O	9.3 ml	8.8 ml	8.2 ml	6.5 ml	4.5 ml	2.4 ml	0.1 ml
10% SDS[a]	0.3 ml	0,3 ml	0.3 ml	0.3 ml	0.3 ml	0.3 ml	0.3 ml
10% APS [a]	0.1 ml	0.1 ml	0.1 ml	0.1 ml	0.1 ml	0.1 ml	0.1 ml
TEMED[a]	0.01 ml	0.01 ml	0.01 ml	0.01 ml	0.01 ml	0.01 ml	0.01 ml

[a] Degas before adding SDS, APS and TEMED

according to size. This is best done by discontinuous SDS poly-acrylamide gel electrophoresis (SDS-PAGE), introduced in 1970 by Laemmli (Fig. 7.4).

Hereby the proteins, per length unit, become equally negatively charged by SDS, migrating towards the positive electrode when subjected to an electrical field. This migration depends entirely on the size of the protein since its own charge is masked by the charge provided by SDS. The change in pH at the borderline between stacking gel and separating gel additionally focusses the protein bands. In a separate lane proteins of known molecular masses serve as molecular markers. The migration distance can be plotted against the logarithm of the molecular mass giving a linear curve in the range in which the gel separates

Table 7.3. Components and volumes for preparation of SDS-polyacrylamide stacking gels

Component	Volume
30% Acrylamide	1.67 ml
1% Bis-acrylamide	1.30 ml
1M Tris/HCl pH 6.8	1.25 ml
H_2O	5.60 ml
10% SDS[a]	0.10 ml
10% APS[a]	0.05 ml
TEMED[a]	0.005 ml

[a] Degas before adding SDS, APS and TEMED.

the proteins efficiently. The range of efficient separation thereby depends on the mesh size of the gel. The mesh size is determined by the ratio of acrylamide vs N,N'-methylene-bis-acrylamide (Tables 7.2, 7.3). High concentrations of acrylamide are used when small proteins are to be analyzed; low concentrations of acrylamide are used when high molecular weight proteins are to be analyzed. When the gel run is finished, proteins can be visualized either by Western blot (see Sect. 7.4) or by staining the gel with Coomassie blue. Coomassie blue stains proteins irreversibly by reacting with arginine residues in the polypeptide chain.

Materials

- Glass plates (approximately 20 cm × 20 cm)
- Set of spacers (thickness: 1 mm)
- Comb (thickness: 1 mm)
- Power supply
- Clamps
- Solutions for preparation of the gel (Tables 7.2, 7.3)
- 1.5 % Agarose solution (for sealing gel)

Buffers and solutions

- Laemmli sample buffer (1×): 10 mM Tris-HCl, pH 6.8; 10 % glycerol; 5 % 2-mercapto-ethanol; 3 % SDS; 0.02 % bromophenol blue
- Electrophoresis buffer: 54 mM Tris-HCl, pH 8.5; 380 mM glycine; 0.1 % SDS
- Coomassie blue solution: 25 % methanol; 12.5 % TCA; 0.1 % Coomassie brilliant blue; R-250
- Destain solution: 7.5 % acetic acid; 5 % methanol

1. Clean glass plates with a detergent followed by alcohol.

2. Assemble the gel support by placing three spacers (two at each side, one at the bottom) between the two glass plates to form a pocket (Fig. 7.4).

3. Fix the sandwich with clamps.

4. Seal the sides holding the spacers with 1.5 % agarose.

5. For the separating gel prepare the polyacrylamide solution according to Table 7.2 depending on the desired gel concentration.

6. Immediately after adding APS and TEMED pour the solution into the assembled gel support leaving a space slightly wider than the length of the slots as represented by the teeth of the comb used to prepare the slots.

7. Carefully overlay the gel solution with distilled water. This allows the gel to polymerize with an even surface and under exclusion of oxygen.

8. Allow to polymerize for at least 1 h at room temperature.

9. Meanwhile, pipet together the components for the stacking gel as described in Table 7.3.

10. A faint line becomes visible between the separating gel's surface and the water when the gel is polymerized.

11. Pour out the water and dry the gel surface with a clean paper tissue.

12. Immediately before casting the stacking gel add APS and TEMED.

13. Cast the stacking gel by filling the remaining space between the glass plates with the stacking gel solution.

14. Immediately put in the comb; use either a normal comb that allows the formation of single slots or a preparative comb that forms one large slot with adjacent single slots for the molecular weight markers (Fig. 7.4). Thus, while the stacking gel is polymerizing the slots are formed.

15. Again, allow to polymerize for at least 1 h. We recommend marking the line between separating gel and stacking gel with a water-resistant felt pen on the glass plate. This later on marks the point when the samples enter the separating gel.

16. When the gel is completely polymerized carefully remove the comb (avoid disrupting the fragile teeth forming the slots) and the bottom spacer.

17. Put the gel in an electrophoresis chamber and fill the chamber with electrophoresis buffer.

18. With a syringe and a bent needle remove any air bubbles trapped in the space created by the bottom spacer (this is important for an even flow of current through the gel). Rinse the slots twice with electrophoresis buffer.

19. Prepare your samples by adding an equal volume of twofold Laemmli sample buffer and boiling the sample for 5–10 min. Keep in mind that more than 150 µg protein/lane cannot be separated efficiently. Also keep in mind the volume of the sample: depending on the size and thickness of the gel a total of 20–100 µl/slot can be loaded. Along with your samples run a sample with proteins of defined sizes as a marker of molecular weights in a separate lane.

20. Run the gel in the following manner: the negative electrode is at the top, the positive electrode is at the bottom. Thus, the current flows "from minus to plus". Start with 75–100 V until the samples enter the separating gel. This is indicated by the bromophenol blue front crossing the line drawn with a felt pen on the glass plates (see step 15).

21. At this point raise the voltage to 180 V (max. 200 V).

22. When the bromophenol blue front reaches the bottom of the gel or starts to leave the gel, the electrophoresis is completed. The gel can now either be stained by Coomassie brilliant blue or further processed for Western blotting.

23. Staining with Coomassie blue: rock the gel gently in Coomassie blue solution for 30 min to 1 h at room temperature. Immerse the gel in destain solution for more than 1 h, frequently changing the destain solution (incubation at 70 °C speeds up the destaining procedure).

7.4
Western Blot

The principle of Western blotting was developed by Towbin et al. (1979). Proteins are separated by SDS-PAGE, transferred to a nitrocellulose membrane and visualized by an antibody followed by a secondary, enzyme-coupled antibody (Figs. 7.5, 7.6). This "two-antibody sandwich" results in amplification of the signal, thereby greatly enhancing the sensitivity of the system. Protein concentrations as low as 0.1 ng can be readily detected. This sen-

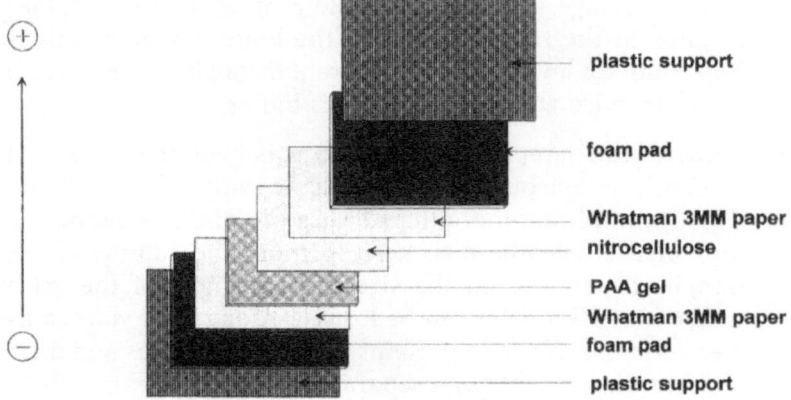

Fig. 7.5. Electrophoretic transfer of proteins to nitrocellulose for Western blot. To transfer proteins electrophoretically to nitrocellulose prepare a sandwich as shown. The components are as follows: plastic support, foam pad, Whatman 3MM paper, gel, nitrocellulose, Whatman 3MM paper, foam pad, plastic support. When preparing the sandwich, carefully remove any air bubbles trapped between the layers to avoid uneven transfer of the proteins

sitivity certainly depends on the quality of the primary antibody. A crucial step in performing a Western blot is the blocking of nonspecific binding sites on the nitrocellulose membrane. There are numerous methods cited in the literature, such as blocking with a solution of TBS/5 % dry milk (fat-free), or TBS/3 % bovine serum albumine (BSA). In our hands, however, the most efficient blocking reagent is TBS/50 % horse serum, which also gives the lowest background. Also, the choice of membrane is an important issue. There are various types of membranes available, including nylon membranes, which are not as brittle as nitrocellulose. Most of these membranes are charged in order to enhance the binding capacity. However, in our experience this results in inhomogeneous binding of some proteins. Therefore, the membrane of choice still is nitrocellulose. Although nitrocellulose with a mesh size of 0.4 μm is appropriate for most purposes, some proteins require membranes with a mesh size of 0.2 μm in order to avoid "blotting through the membrane". The scleroderma-specific CENP-B antigen, a protein with a molecular mass of 80 kDa and associated with the centromere, is detectable only when using short transfer times (approximately 2–3 h) and nitrocellulose membranes with a mesh size of 0.2 μm. For detection of autoantigens we use subcellular fractions prepared

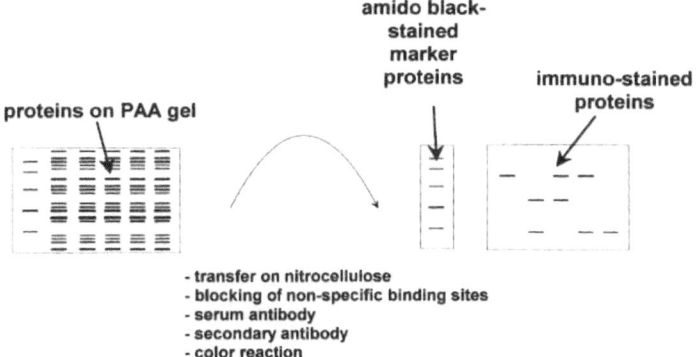

Fig. 7.6. Principle of Western blot. Proteins are separated on a polyacrylamide gel and transferred to nitrocellulose (see Fig. 7.5). After transfer proteins are reversibly stained with Ponceau S red. Cut off the lane with marker
proteins and stain with amido black. On the remaining nitrocellulose nonspecific binding sites are blocked with TBS/50 % horse serum. The nitrocellulose is incubated with the serum dilution and bound antibodies are
detected by incubation with a secondary, enzyme-coupled anti-antibody
followed by a color reaction with the appropriate substrate. The substrate
forms an insoluble precipitate on the nitrocellulose thus indicating the
position of the antigen

according to the protocols described in Section 7.2. The CENP-B
and CENP-A antigens, both protein components of the centromere, are detectable in the soluble nuclear fraction, the snRNPs,
also known as Sm and U1 RNP antigen along with Ro and La
(SSA, SSB), are found in the RNP fraction. The nucleolar autoantigens fibrillarin and PM/Scl are detected in the nucleolar fraction. These are only examples of some of the most common
autoantigens. When dealing with an autoimmune serum of
unknown specificity care must be taken to establish the appropriate blotting conditions and substrates.

Materials

– Transfer chamber
– Power supply
– Plastic supports
– Foam pads
– Whatman 3MM filter paper
– Nitrocellulose (mesh size: 0.2–0.4 μm)

Buffers and solutions

– Transfer buffer: 25 mM Tris-HCl, pH 8.6; 192 mM glycine; 20 % methanol
– TBS: 10 mM Tris-HCl, pH 7.6; 150 mM NaCl
– Ponceau S red: 0.2 % Ponceau S; 3 % TCA
– Amido black: 10 % acetic acid; 45 % methanol; 0.1 % amido black
– Destain solution: 7.5 % acetic acid; 5 % methanol
– AP buffer: 100 mM Tris-HCl, pH 9.5; 100 mM NaCl; 5 mM MgCl$_2$
– Stop buffer: 20 mM Tris-HCl, pH 8.0; 5 mM EDTA
– Substrate stock solutions: NBT (4-nitroblue tetrazoliumchloride), 50 mg/ml in 70 % di-methyl-formamide; BCIP(5-bromo-4-chloro-3-indoxyl phosphate), 50 mg/ml in 100 % di-methyl-formamide. Prepare the substrate by mixing 66 µl NBT stock solution and 66 µl BCIP stock solution with 10 ml AP buffer.

Procedure

Transfer of Proteins to Nitrocellulose

1. Cut Whatman 3MM filter paper and nitrocellulose to the desired size; i.e., so that the Whatman 3MM filter paper is slightly larger than the gel and the nitrocellulose is the same size as the gel.

2. Fill a tray with transfer buffer so that the transfer sandwich to assemble in the following step is soaked, but not completely immersed.

3. Assemble the transfer sandwich as depicted in Fig. 7.5: place the plastic support in the tray, soak a foam pad and place it on the plastic support thereby carefully removing any air bubbles trapped in the squamous material of the foam pad (air bubbles interfere with transfer of the proteins). A soaked filter paper is placed on the foam pad. Rinse the filter paper with a few drops of transfer buffer and roll a Pasteur pipette evenly over the filter paper to remove any trapped air bubbles. Cut off the stacking gel with a scalpel and place the separating gel on the filter paper, again smoothening it with a Pasteur pipette. Prerinse the nitrocellulose with transfer buffer, fit it exactly on the gel and remove air bubbles as described. Cover it with a rinsed filter paper, the second foam pad and the second plastic support. Place the whole sandwich in the transfer chamber filled with transfer buffer. Keep in mind the orienta-

tion of the sandwich: the current later on has to flow from the negative electrode to the positive electrode.

4. Transfer overnight at 50 V and 4 °C, thereby stiring the buffer continuously with a magnetic stirer and a stirring bar to avoid a temperature gradient.

5. The following morning raise the voltage to 70 V for 30 min to complete the transfer efficiently.

Immunostaining of the Transferred Proteins

1. Dismantle the sandwich keeping in mind the orientation of the nitrocellulose (proteins are now on the side of the nitrocellulose facing the gel. The following steps are performed with the nitrocellulose "face-up").

2. Rinse the nitrocellulose briefly in TBS.

3. Stain for 10 min at room temperature with Ponceau S red.

4. Destain with TBS thereby localizing the marker proteins.

5. Cut off the lane with the marker proteins using a ruler and a scalpel and stain separately with amido black (completely destain with TBS, stain in amido black for 10 min and destain with destain buffer, air-dry).

6. Completely destain the remaining nitrocellulose in TBS and place in blocking buffer.

7. Rock for 2 h at room temperature.

8. Discard the blocking buffer, rinse the nitrocellulose briefly in TBS and incubate in the serum diluted in TBS/3 % horse serum (a good dilution is 1:100) for 90 min.

9. Wash the nitrocellulose with three washes with TBS/0.1 % Tween-20 for 10 min, followed by three washes with TBS for 10 min.

10. Incubate in a 1:5000 dilution of the secondary antibody (alkaline phosphatase-coupled goat anti-human IgG, DIANOVA, Hamburg, Germany) for 90 min at room temperature.

11. Wash as described in step 9.

12. Briefly equilibrate the nitrocellulose in AP buffer while preparing the substrate in AP buffer.

13. Discard the equilibration solution and replace with the substrate. Perform the color reaction observing development of the purple precipitate. The color reaction should take between 1 and 5 min.

14. Stop by discarding the substrate and incubate in stop solution for 10 min. The nitrocellulose can then be photographed for documentation and air-dried for storage.

7.5
Affinity Purification of Autoantibodies

When dealing with autoimmune sera containing a spectrum of autoantibodies, it is often neccessary to dissect this multispecific immune response into its single components in order to analyze defined antibody/antigen systems. This is done by affinity purification of antibodies contained in the serum (Fig. 7.7). The antigen, either in a crude mixture of proteins or already purified (e.g., a recombinant protein), is separated on a preparative polyacrylamide gel and transferred to nitrocellulose (see Sect. 7.4). The antigen then is localized either by immunostaining of adjacent strips, when contained in a mixture of proteins, or simply by staining with Ponceau S red, when it is already purified. The area containing the antigen is excised. The antibodies are allowed to react with their corresponding antigen, unbound antibodies are washed off the complex followed by elution of specifically bound antibodies. Depending on the intended use of the affinity-purified antibodies one can either use the eluate directly or concentrate it. The affinity-purified antibodies can be re-incubated on nitrocellulose strips (for Western blot) or, in concentrated form, used to perform indirect cytoimmunofluorescence. We even used them to screen an expression library (Blüthner and Bautz 1992). In our lab we employ two methods of affinity purification of antibodies: elution by low pH (Smith and Fisher 1984) or elution by KSCN (Olmstedt 1981; Krohne et al. 1982). Both methods work equally well. However, the eluted antibodies are stable for a limited time only (approximately 2–3 days at 4 °C).

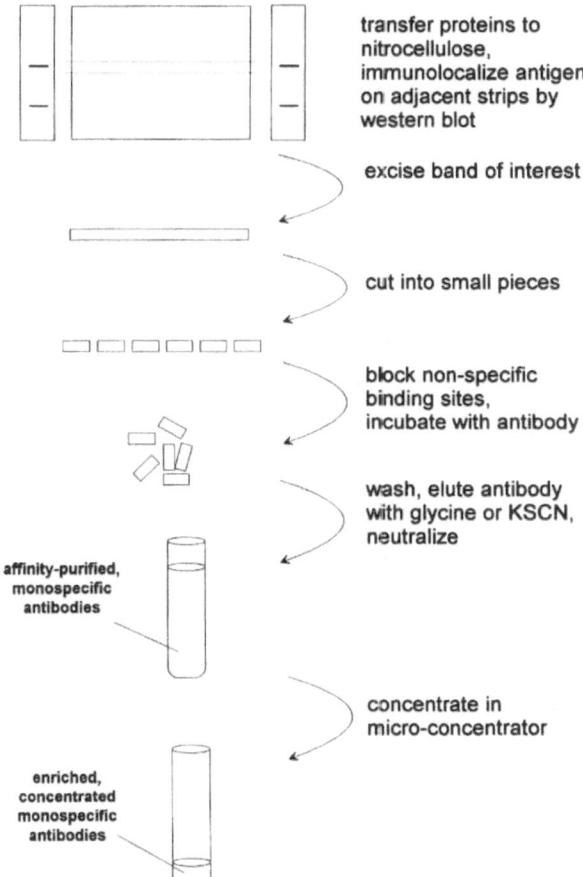

transfer proteins to
nitrocellulose,
immunolocalize antigen
on adjacent strips by
western blot

excise band of interest

cut into small pieces

block non-specific
binding sites,
incubate with antibody

wash, elute antibody
with glycine or KSCN,
neutralize

affinity-purified,
monospecific
antibodies

concentrate in
micro-concentrator

enriched,
concentrated
monospecific
antibodies

Fig. 7.7. Affinity purification of antibodies. Proteins separated on a preparative SDS-PAGE gel are transferred to nitrocellulose. To localize the protein of interest adjacent strips are cut off and a Western blot is performed. The band of interest is excised and cut into small pieces. The small strips are blocked with TBS/50 % horse serum followed by incubation with the serum dilution. After washing bound antibodies are eluted either by glycine or by KSCN as described, concentrated and are then ready to use. Affinity-purified antibodies are stable for only a limited time at 4 °C and must immediately be further processed

Materials

– Screw-cap tubes (15 ml)
– Centricon-30 microconcentrators
– Additional materials: see Western blot (Sect. 7.4)

Buffers and
solutions
- TBS: 10 mM Tris-HCl, pH 7.6; 150 mM NaCl
- PBS: 8 mM Na_2HPO_4; 1.5 mM $KHPO_4$; 140 mM NaCl; 2.6 mM KCl. Adjust pH to 7.3
- Glycine-elution buffer: 5 mM glycine; 150 mM NaCl; 0.1 % Tween-20. Adjust pH to 2.4
- Neutralization buffer: 1 M Tris-HCl, pH 8.0
- KSCN solution: 3 M KSCN

Procedure

Localizing and Isolating the Antigen of Interest

1. Run a preparative SDS-PAA gel as described above (i.e., prepare the gel with a large slot instead of several single slots; see Fig. 7.4B).

2. Transfer the proteins to nitrocellulose as described in "Transfer of Proteins to Nitrocellulose" (Sect. 7.4).

3. Stain with Ponceau S red as described above.

4. With a scalpel cut off the molecular markers and a strip of nitrocellulose at the very right and left sides containing transferred proteins as depicted in Fig. 7.7. Store the remaining nitrocellulose at 4 °C in TBS.

5. Localize the antigen of interest by immunostaining as described in "Immunostaining of the Transferred Proteins" (Sect. 7.4).

6. Cut out a horizontal strip from the remaining nitrocellulose representing the relative position of the antigen of interest, cut this strip into small pieces and place the small strips into a screw-cap tube.

7. Block nonspecific binding sites by incubation in TBS/50 % horse serum for 2 h at room temperature.

8. Discard the blocking solution and incubate the strips in a 1:50 serum dilution.

9. Incubate for 90 min at room temperature on a rocker.

10. Wash as described in step 9 of "Immunostaining of the Transferred Proteins".

11. Discard washing solution.

Elution of Bound Antibodies

1. Place the strips in a screw-cap tube and add 1.5 ml Elution by low pH glycine elution buffer.

2. Rock for 10 min at room temperature, occasionally vortex briefly.

3. Meanwhile prepare a second tube with 0.6 ml neutralization buffer.

4. When the elution is finished aspirate the elution buffer with a Pasteur pipette and add it to the second tube with the neutralization buffer (avoid keeping the pH labile eluted antibodies unbuffered).

5. Repeat steps 1–4.

6. If you wish to concentrate your eluted antibodies, see below ("Concentration and Dialysis of Eluted Antibodies").

Elution with KSCN

1. Place the strips in a screw-cap tube and add 1.0 ml 3 M KSCN.

2. Rock gently for 5 min at room temperature.

3. Aspirate the eluted antibodies with a Pasteur pipette and immidiately add to a second tube containing 2 ml PBS thus diluting KSCN from 3 to 1 M.

4. To concentrate and dialyze the antibodies, see below ("Concentration and Dialysis of Eluted Antibodies").

Concentration and Dialysis of Eluted Antibodies

1. Fill the eluted antibodies in Centricon-30 microconcentrators and place in a centrifuge.

2. Centrifuge for 20 min at $1500 g$ at 4 °C; this concentrates the solution to approximately 200 µl. If the membrane is clotted, resulting in inefficient concentration of the solution, extend the centrifugation time.

3. Fill the microconcentrator with PBS (the container holds approximately 2 ml) and recentrifuge as before.

4. Repeat step 3.

5. Place the cap on top and turn the concentrator.

6. Centrifuge briefly (approximately 2 min) at 1500 g thus saving the concentrated antibody solution.

7. Use the concentrated antibody undiluted in indirect cytoimmunofluorescence or in a dilution of 1:50 in Western blot experiments.

7.6
ELISA with Recombinant Autoantigens

When a purified autoantigen is available (e.g., recombinant antigen) and many sera are to be tested, the method of choice is certainly the ELISA, as this system is easy to handle and saves material (i.e., serum). The principle of the ELISA is depicted in Fig. 7.8. Wells of a polystyrene plate are coated with the antigen, followed by a blocking step to block non-specific binding sites. The antigen is then detected with the serum antibody and a sec-

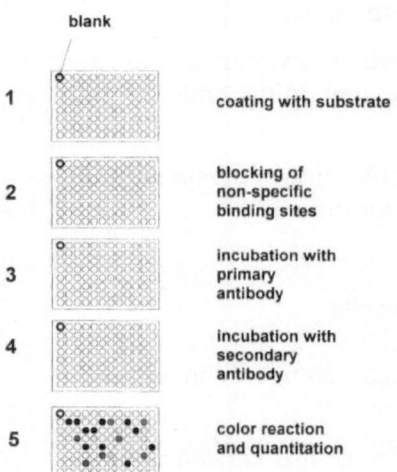

Fig. 7.8. Principle of ELISA. Wells of a 96-well polystyrene plate are coated with antigen. After washing nonspecific binding sites are blocked with PBS/0.2 % Tween-20. Plates are incubated with the serum dilution and washed. Bound antibodies are detected by incubation with a secondary, alkaline phosphatase-coupled, anti-antibody followed by a color reaction with the substrate. The product of this color reaction is soluble allowing quantitation of the reaction with an ELISA reader (which essentially is a spectrophotometer). The intensity of the color reaction is a measure of the amount of bound antibody

ondary, enzyme-coupled anti-antibody followed by a color reaction. The recombinant antigens we use in our lab are derived either from the maltose binding protein system (vector pMAL) or from the histidine tag system (vector pDS). Both systems allow purification of the recombinant antigen over an affinity column. Following the affinity purification we usually purify the samples by gel filtration to get rid of any remaining bacterial contamination. Purity of the recombinant antigen is essential when screening human sera since most sera contain antibodies against *E.coli* proteins, potentially leading to false positive results in the assay. Check your antigen preparations by PAGE prior to running the assay. Also run pilot assays to determine the appropriate conditions (antigen concentration, serum dilution).

Materials

- ELISA plates (96-well plates)
- ELISA reader
- Secondary antibody (e.g., alkaline phosphatase-coupled goat anti-human IgG)
- pNPP (*p*-nitrophenyl phosphate)

- Coating buffer (stock solution): 100 mM Na_2CO_3; 100 mM NaHCO$_3$. Adjust Na_2CO_3 solution to pH 9.5 with NaHCO$_3$. For the working solution, dilute to 50 mM with H_2O **Buffers and solutions**
- PBS: 8 mM Na_2HPO_4; 1.5 mM KH_2PO_4; 140 mM NaCl; 2.6 mM KCl. Adjust pH to 7.3
- Block solution: PBS/0.2 % Tween-20
- Wash buffer: PBS/0.05 % Tween-20
- Substrate buffer (always freshly prepare the substrate buffer): 10 mM diethanolamine; 0.5 mM $MgCl_2$. Adjust to pH 9.5 with 1 M HCl. Prior to use dissolve one tablet pNPP in 5 ml substrate buffer
- Stop solution: 100 mM EDTA. Adjust to pH 7.5 with 1 M NaOH

Procedure

1. Resuspend the antigen in coating buffer. When resuspending the antigen take into acount the binding capacity of the plastic material. A single well, under the conditions described here, binds approximately 200 ng. Therefore, a good concentration to start with is 2 µg/ml or 200 ng/100 µl.

2. Coat the plate by pipetting 100 µl of your antigen solution into the ELISA wells. The first well (or the first row, depending on your ELISA reader) is filled with 100 µl coating buffer without antigen. This well or row serves as the blank.

3. Incubate either overnight at 4 °C or 2 h at room temperature. When incubating overnight the plate should be kept at room temperature for at least 30 min the following morning. This avoids a temperature gradient within the plate which could lead to inconsistent binding in the following steps.

4. Remove the remaining antigen solution by tapping the plate vigorously on a stack of tissue papers.

5. Wash the wells with wash buffer. This is simply done by filling the wells with wash buffer using a 500 ml or 1000 ml squirt bottle. Leave the wash buffer for 5 min. Pour out as before. Repeat this procedure twice. Repeat another three times using PBS instead of PBS/0.05 % Tween-20.

6. Fill the wells with blocking solution (approximately 350 µl/ well). Leave at room temperature for 2 h. Nonspecific protein binding sites are blocked in this step.

7. Meanwhile prepare your serum dilutions. We routinely dilute our sera 1:100 in PBS/0.2 % Tween-20. The dilution factor depends on the titer of the serum, the affinity of the antibody and the antigen concentration and has to be determined individually for each serum.

8. Remove the block solution as in step 4.

9. Fill the wells with 100 µl of the serum dilution in blocking solution.

10. Leave at room temperature for at least 90 min.

11. Remove the serum dilution and wash as described in steps 4 and 5.

12. Fill the wells with 100 µl of secondary antibody diluted in blocking solution. We use an alkaline phosphatase-coupled F(ab') fragment of goat anti human IgG diluted 1:5000 from DIANOVA, Hamburg, Germany. Other companies offer comparable antibodies which have to be diluted differently (refer to the data sheet of your specific secondary antibody).

13. Incubate for at least 90 min at room temperature.

14. Remove the antibody solution and wash as described in steps 4 and 5.

15. Equlibrate the wells twice with 100 µl substrate buffer per well.

16. Remove the substrate buffer and add 100 µl substrate solution per well.

17. Incubate the plate for 20–30 min at room temperature in the dark (simply cover the plate with a cardboard box).

18. Stop by adding 100 µl/well stop solution.

19. Measure the OD at 405 nm in an ELISA reader against your blank.

References

Blüthner M, Bautz FA(1992) Cloning and characterization of the cDNA coding for a polymyosis-scleroderma overlap syndrome-related nucleolar 100 kDa protein. J Exp Med 176: 973–980

Gey GD, Coffman WD, Kubicek MD (1952) Tissue culture studies of the proliferative capacity of cervical carcinoma and normal epithelium. Cancer Res 12: 264–265

Guldner HH, Lakomek HJ, Bautz FA (1983) Identification of human Sm and (U1) RNP antigens by immunoblotting. J Immunol Meth 64:45–59

Kalckar HM (1947) Differential spectrophotometrie of purine compounds by means of specific enzymes. III. Studies of the enzymes of purine metabolism. J Biol Chem 167:461–475

Krohne GR, Stick R, Kleinschmidt J, Moll R, Franke WW, Hausen P (1982) Immunological localization of a major karyoskeletal protein in nucleoli of oocytes and somatic cells of Xenopus laevis. J Cell Biol 94:749–754

Laemmli UK (1970) Cleavage of structural proteins during the assembly of the head of bacteriophage T4. Nature 227:680–685

Olmsted JB (1981) Affinity purification of antibodies from diazotized paper blots of heterogeneous protein samples. J.Biol.Chem. 256:11955–11957

Smith DE, Fisher PA (1984) Identification, developmental regulation, and response to heat shock of two antigenically related forms of a major nuclear envelope protein in Drosophila embryos: application of an improved method for affinity purification of antibodies using polypeptides immobilized on nitrocellulose blots. J Cell Biol 99:20–28

Tan EM (1989) Antinuclear antibodies: Diagnostic markers for autoimmune diseases and probes for cell biology. Adv.Immunol 44:93–151

Towbin H, Staehelin T, Gordon J (1979) Electrophoretic transfer of proteins from polyacrylamide gels to nitrocellulose sheets: procedure and some applications. Proc Natl Acad Sci USA 76:4350–4354

van Venrooij WJ, Maini RN (1994) Manual of biological markers of disease. Kluwer Academic, Dordrecht

Chapter 8

In Vitro Splicing of Pre-mRNA in HeLa Extracts

Johannes Schenkel[1], Frank Jung[2], Apostolia Guialis[3], and Angela Krämer[4*]

Introduction

The protein-coding sequences in most messenger RNA precursors (pre-mRNAs) of higher eukaryotes are interrupted by intervening sequences (or introns), which are removed from the primary transcript in a process termed splicing (for review see Moore et al. 1993; Krämer 1996). The reaction occurs in the nucleus either co- or post-transcriptionally. The coding sequences (or exons) are joined and the mature mRNA is exported to the cytoplasm where it is translated into protein. Intron removal is an essential process because introns usually have no or only little protein-coding potential and the failure to remove them from the mRNA can introduce in-frame stop codons resulting in premature termination of translation and thus non-functional proteins. Many pre-mRNAs contain more than one intron, which provides a means for the regulation of gene expression by alternative splicing events. The selection of alternative 5' and 3' splice sites or the inclusion/exclusion of entire exons can result in an on/off switch of a particular gene or in the translation of different protein isoforms (often with specific functions) from one and the same pre-mRNA, thus augmenting the coding repertoire of the genome.

* Corresponding author: Angela Krämer: Tel.: (+41)-22–7026750;
  Fax: (+41)-22–7026750; e-mail: KRAEMER@CELLBIO.UNIGE.CH
[1] II. Institute of Physiology, University of Heidelberg, 69120 Heidelberg, Germany
[2] Institute of Molecular Genetics, University of Heidelberg, Im Neuenheimer Feld 230, 69120 Heidelberg, Germany
[3] The National Hellenic Research Foundation, Institute of Biological Research and Biotechnology, Athens, Greece
[4] Department of Cell Biology, University of Geneva, 30, Quai Ernest-Ansermet, 1211 Geneva 4, Switzerland

After the discovery of split genes in vivo methods were employed to demonstrate the importance of splicing and introns for efficient gene expression and *cis* elements required for accurate splicing were defined. However, the splicing reaction mechanism was only unraveled after in vitro systems that accurately mimic the in vivo situation became available. The most commonly used system employs nuclear extracts from HeLa suspension cells (Dignam et al. 1983; Dignam 1990), however extracts from other cell types and organisms were also developed (Eperon and Krainer 1994) and methods for the preparation of extracts from small quantities of cells (either suspension cells or cells growing in monolayers) have been described (Krämer and Keller 1990; Lee and Green 1990). In addition, extracts can be prepared from yeast cells and studies using yeast extracts have complemented data obtained with genetic approaches in this organism (Newman 1994; Beggs 1995).

Initial in vitro studies used purified nuclear pre-mRNAs as substrates and the products of the in vitro splicing reaction were analyzed by primer extension and S1-nuclease mapping. These approaches had the disadvantage that only a subset of the RNA products generated during the reaction could be detected. When in vitro methods to transcribe uniformly labeled pre-mRNAs from cloned templates with bacteriophage RNA polymerases were developed (Green et al. 1983; Melton et al. 1984), all products of the splicing reaction could be visualized in denaturing polyacrylamide gels and it became apparent that the removal of introns from the pre-mRNA proceeds in two well-defined biochemical steps (Moore et al. 1993). In vitro transcription also made it possible to generate mutant transcripts and the *cis*-requirements for splice site selection were defined in detail. Thus, in vitro systems have been valuable tools for the identification and characterization of the components that are essential for the splicing process; fractionating splicing extracts have facilitated the purification of splicing components and cloning of the corresponding cDNAs.

The major characteristics of the splicing reaction in higher eukaryotes can be summarized as follows (for review see Moore et al. 1993; Krämer 1996; Fig. 8.1). Three main *cis*-acting elements have been identified that are required for the efficient and accurate removal of introns. The 5' splice site is defined by about six nucleotides of the consensus sequence GURAGU (where denotes the splice site; A = adenosine, C = cytosine, G = guanosine, U = uridine, R = purine, Y = pyrimidine, N = any

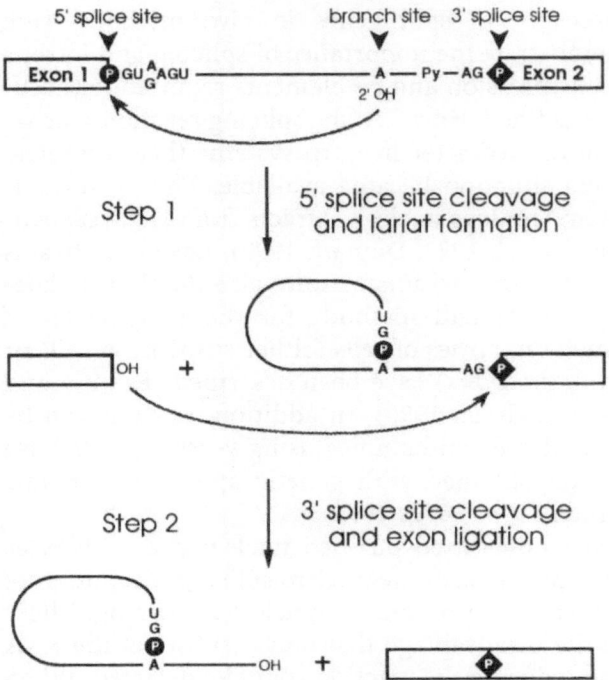

Fig. 8.1. The catalytic steps of splicing. Exons are shown as *boxes*, the intron as a *line*. Conserved intron sequences are indicated and the phosphates at the splice sites are shown as *circles* and *squares*. The *dashed arrows* indicate the nucleophilic attack of the hydroxyl groups on the splice junctions. (Modified from Krämer 1996)

nucleotide), and the 3' splice site is defined by YAG, which is in many cases preceded by a polypyrimidine tract. A third sequence important for splicing, the branch site, is located at a distance of 20–40 nucleotides upstream of the 3' splice site; the less well-defined consensus sequence is YNYURAY.

The catalysis of splicing involves two consecutive transesterification reactions which can be followed by electrophoresis of the reaction products in denaturing polyacrylamide gels. In a first step the branch site adenosine attacks the 5' splice site leading to cleavage of the phosphodiester bond at this site and the concomitant covalent linkage of the 5' end of the intron to the branch site adenosine in an unusual 2'5'-phosphodiester bond (Fig. 8.1). The intermediates of the reaction are the cleaved-off 5' exon and the intron-3' exon in a branched circular form or lariat. The second step involves the nucleophilic attack of the 3'

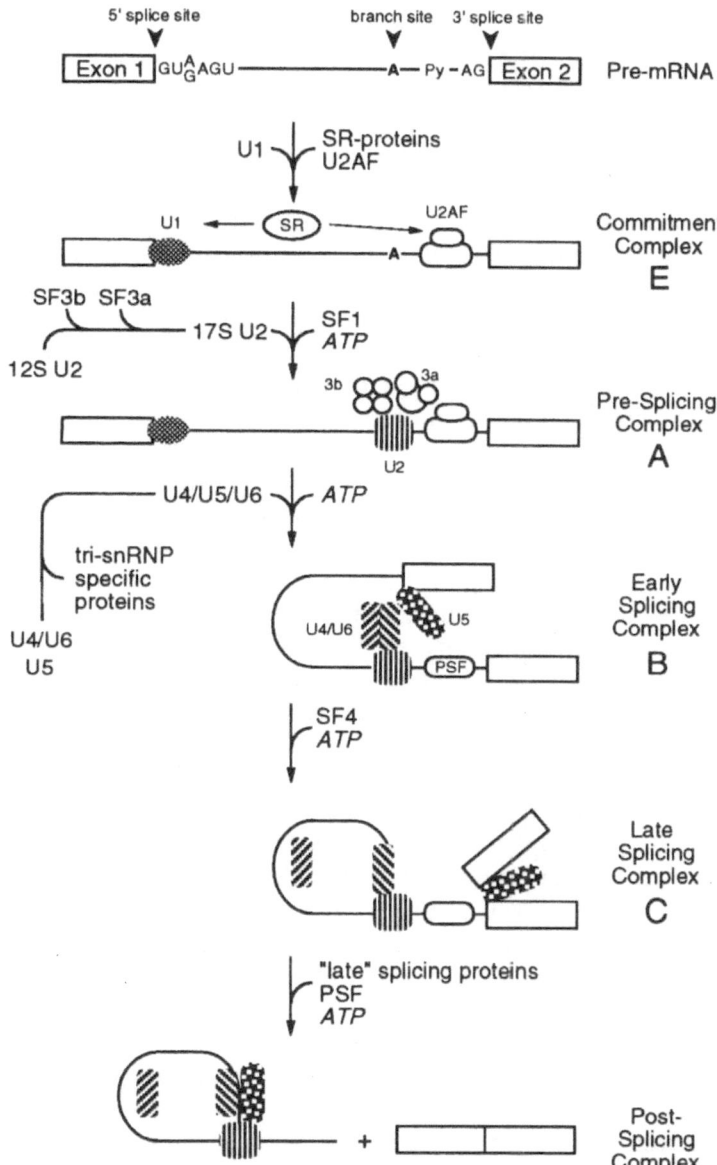

Fig. 8.2. The spliceosome assembly pathway. SnRNPs are indicated by *filled symbols*, protein factors by *open symbols*. For details see text and Krämer (1996). (Modified from Krämer 1996)

hydroxyl group of the 5' exon at the phosphodiester bond at the 3' splice site. At the same time that cleavage occurs, the exons are ligated and the intron is released in the form of a lariat. The 2'5'-phosphodiester bond at the branch site is cleaved by a debranching enzyme followed by degradation of the intron.

The in vitro splicing reaction requires incubation at 30 °C, low concentrations of monovalent and divalent cations and ATP as an energy source. Depending on the pre-mRNA substrate a lag-phase in the appearance of the intermediates of 20–30 min is observed and the final reaction products are usually seen after 30–40 min. During this lag phase the pre-mRNA is assembled into a splicing-competent structure which has been termed the spliceosome. The formation of the spliceosome occurs in several steps and requires small nuclear ribonucleoprotein particles (snRNPs) as well as non-snRNP splicing proteins which interact with one another and with the consensus sequences in the pre-mRNA (Fig. 8.2). Initially, splicing complexes were identified by sedimentation of splicing reactions in glycerol or sucrose gradients, which demonstrated that spliceosomes are complex structures comparable in size to the ribosomal subunits (30S–60S). Later, methods to separate spliceosomes by native polyacrylamide gel electrophoresis were developed which provides a more convenient and faster approach and allows the handling of many reactions at the same time.

The first complex that can be detected by native gel electrophoresis is complex H. This complex forms in the absence of any intronic sequences and its formation does not require either incubation at elevated temperature or ATP. Thus, this complex is not specific for the splicing reaction and it is thought to represent the binding of general RNA-binding proteins to the pre-mRNA. The first splicing-specific complex (complex E or commitment complex) is formed by binding of U1 snRNP to the 5' splice site and recognition of the 3' splice site polypyrimidine tract by a protein factor (Fig. 8.2). This complex is usually not resolved in native polyacrylamide gels; however, it can be detected by chromatographic methods. In the following step, which requires ATP, U2 snRNP binds to the branch site thereby generating pre-splicing complex A which forms in vitro within 2–5 min of incubation at 30 °C and is readily detected in native polyacrylamide gels. A triple-snRNP consisting of U4/U6 and U5 snRNPs then binds to complex A and splicing complex B is formed within 10–15 min of incubation. A conformational change results in splicing complex C in which the catalysis of splicing takes place. This complex is relatively unstable in native polyacrylamide gels and cannot always be resolved from complex B. After the splicing reaction is completed, the mature mRNA is released from the spliceosome. The intron remains associated (for a certain time) with splicing components and can be detected at late times of the splicing reaction as a complex that migrates between complexes A and B/C.

In addition to the snRNPs, many protein factors participate in the splicing reaction. A description of their function has been omitted here for reasons of brevity. For review the reader is referred to articles by Moore et al. (1993) and Krämer (1996).

8.1
Preparation of HeLa Cell Nuclear Extracts

As outlined above, the splicing of pre-mRNA requires many nuclear components, such as snRNPs and protein factors. These are present in nuclear extracts that can conveniently be prepared from HeLa cells grown in suspension cultures. Detailed descriptions of HeLa cell culture have been published by Krämer and Keller (1990) and Eperon and Krainer (1994). The procedure for the preparation of nuclear extracts is based on the method developed by Dignam et al. (1983). For experimental details of this method the reader is referred to the article by Dignam (1990). Extracts are preferably prepared from freshly harvested cells. Extracts prepared from cells frozen as a pellet can also be used; however, the efficiency of splicing may be slightly reduced.

For the following procedure we routinely use cells harvested from 15 l of suspension culture at a density of $5-10 \times 10^5$ cells/ml. The method can be scaled up or down. Extracts of small amounts of cells (for example 3×10^7 cells) can be prepared by the method of Lee and Green (1990).

The preparation of HeLa cell nuclear extracts involves hypotonic swelling of the cells. Nuclei are collected after disruption of the plasma membrane. Nuclear components are extracted and chromatin is precipitated by addition of high salt. After removal of debris the nuclear extract is dialyzed against a suitable buffer. The cytoplasmic fraction of the cells can be used to prepare a cytoplasmic S100. These extracts are deficient in splicing; however, they represent a useful system for the analysis of one type of splicing protein (see Krainer et al. 1990; Zahler et al. 1992).

▓ Materials

- Dialysis membranes
- Glass/glass homogenizer type Dounce
- Sorvall centrifuge with rotors
- Ultracentrifuge with rotors

Reagents and buffers

Note: All buffers used for the preparation of extracts should be prepared with sterile double-distilled H_2O.

- HeLa cells, freshly harvested or frozen
- Buffer A: 10 mM Hepes-KOH, pH 7.9; 1.5 mM $MgCl_2$; 10 mM KCl; 0.5 mM DTT
- Buffer B: 0.3 M Hepes-KOH, pH 7.9; 1.4 M KCl; 0.03 M $MgCl_2$
- Buffer C/low salt: 25 % (v/v) glycerol; 20 mM Hepes-KOH, pH 7.9; 1.5 mM $MgCl_2$; 0.2 mM EDTA; 20 mM NaCl; 0.5 mM DTT; 0.5 mM phenylmethylsulfonyl fluoride (PMSF)
- Buffer C/high salt: same as buffer C/low salt but containing 1.2 M NaCl
- Buffer D: 20 mM Hepes-KOH, pH 7.9; 20 % (v/v) glycerol; 100 mM KCl; 0.2 mM EDTA; 0.5 mM DTT; 0.5 mM PMSF
- DTT: A 0.5 M stock solution can be frozen at $-20\,°C$. Add to buffers immediately before use.
- PMSF: 0.05 M; dissolve in 96 % ethanol. This solution can be stored for a few days at $-20\,°C$. Add to buffers immediately before use.
- PBS

▓ Procedure

Preparation of splicing extracts

Note: Unless stated otherwise, the whole procedure should be performed at 4 °C or on ice using precooled buffers.

1. If frozen cells are used, add PBS and thaw cells **quickly** under lukewarm tap water.

2. Spin for 10 min at 2500 rpm in a Sorvall centrifuge. Discard supernatant.

3. Resuspend cells in PBS and transfer to 50 ml screw-cap tubes. Spin for 10 min at 2500 rpm in a Sorvall centrifuge.

4. Measure the packed cell volume (PCV).

5. Resuspend the cells in PBS (5 volumes PCV). Spin for 10 min at 2500 rpm in a Sorvall centrifuge.

6. Resuspend the cells in buffer A (2 volumes PCV).

7. Leave on ice for 10 min to swell.

8. Homogenize ten times in a Dounce homogenizer.

9. Spin for 15 min at 3000 rpm in a Sorvall centrifuge.

10. Determine the total volume (vol 1) of cells and supernatant.

11. Remove the supernatant and measure its volume (vol 2). This supernatant can be used to prepare a cytoplasmic S100 extract (see step 20).

12. Determine the packed nuclear volume: PNV = (vol 1)−(vol 2).

13. Resuspend the nuclei in buffer C/low salt (0.5 vol PNV) by gently agitating the tube.

14. **Dropwise,** add buffer C/high salt (0.5 vol PNV) while gently agitating the tube.

15. Homogenize ten times in a Dounce homogenizer.

16. Transfer the nuclei into a small glass bottle and place on ice. Extract the nuclei for 30 min with gentle stirring.

17. Spin for 30 min in a Sorvall centrifuge (HB-4 swinging bucket rotor or equivalent) at 16 500 g.

18. Transfer the supernatant to dialysis tubing and dialyze against three changes (1–2 h each) of an ~100-fold excess of buffer D.

19. Spin for 20 min in a Sorvall centrifuge (SS34 or HB4 rotor) at 24 000 g to remove the precipitate that can form during dialysis.

20. If a cytoplasmic S100 extract is desired in addition to the nuclear extract start with the following steps during the 30 min incubation of the nuclei (step 16). Add buffer B (0.11 × [vol 2]) to the supernatant and mix well.

21. Spin for 1 h at 100 000 g in an ultracentrifuge.

22. Transfer the supernatant to dialysis tubing and dialyze as in step 18. Spin as in step 19.

23. Store nuclear extract and cytoplasmic S100 in aliquots at
    −80 °C. Extracts can be thawed and frozen a few times with-
    out considerable loss of activity. Repeated freeze-thawing
    should however be avoided.

8.2
In Vitro Synthesis of the Pre-mRNA Substrate

Splicing reactions are usually performed with simple RNA substrates that contain two exons and a short intron. These substrates are generated by transcription with bacteriophage RNA
polymerases. To simplify the analysis of the reaction products,
the RNA substrate is uniformly labeled during the transcription
reaction. In addition, the RNA is capped at the 5' end by including the cap analogue m^7GpppG in the reaction.

The RNA substrate used here is transcribed with T3 RNA
polymerase from plasmid pBSAL4 linearized with EcoRI.
pBSAL4 (Lamond et al. 1987) contains part of the rabbit β-
globin DNA and consists of the 3' portion of exon 2, intron 2
from which 254 internal nucleotides have been deleted and the 5'
portion of exon 3 (Fig. 8.3). The size of the in vitro synthesized
pre-mRNA is 428 nucleotides (56 nucleotides of exon 2, 319 nucleotides of intron 2 and 53 nucleotides of exon 3). The protocol
given below is based on the procedure of Melton et al (1984). For
additional protocols describing the preparation of pre-mRNA
substrates see Yisraeli and Melton (1989), Krämer and Keller
(1990) and Chabot (1994). Protocols and the application guide
book from Promega provide further details and remarks on the
specific activity of in vitro synthesized RNA.

Materials

Equipment
- Microfuge
- Water bath (37 °C)
- Gel electrophoresis apparatus and power supply
- GF/C glass fiber filters
- Microfuge
- Parafilm
- Plexiglas screen
- Razor blades or scalpel

- Saran wrap
- Scintillation counter
- Sorvall centrifuge with SS34 and HB-4 rotors
- X-ray film and developing equipment

- 1% agarose minigel (see Sect. 8.4, "Preparation of Agarose   **Reagents**
  Minigels")
- pBSAL4 plasmid DNA (0.5 µg/µl)
- Chloroform/isoamyl alcohol: (24:1 v/v)
- *Eco*RI restriction endonuclease (25 U/µl)
- 2× *Eco*RI restriction digestion buffer (usually supplied with
  the enzyme)
- Phenol: saturated in 0.1 M Tris-HCl, pH 8.0
- Sodium acetate: 3 M, pH 5.2
- $[\alpha^{32}P]UTP$: 7.5×10^5 Bq/µl, specific activity 3×10^{13} Bq/nmol
- Crush-and-soak solution: 0.5 M ammonium acetate, pH 7.5;
  10 mM magnesium acetate; 0.1% SDS; 0.1 mM EDTA; 10 mM
  Tris-HCl, pH 7.5
- Denaturing 7% polyacrylamide/7 M urea gel (see Sect. 8.4,
  "Preparation of a Native Polyacrylamide/Agarose Gel")
- DNA template: pBSAL4 DNA linearized with *Eco*RI (0.5 µg/µl)
- DNase (RNase-free): 1 U/µl
- DTT: 0.75 M in H_2O; store at $-20\,°C$
- Ethanol: 96%
- Formamide: deionized
- m^7GpppG cap analogue (Pharmacia 27–4631–01): 10 mM
  unmethylated GpppG (Pharmacia 27–4635–01), which is less
  expensive, can be used instead of m^7GpppG without effect on
  the efficiency or accuracy of splicing in vitro.
- NaCl: 1 M and 5M
- Phenol/chloroform: phenol/chloroform/isoamyl alcohol (25:
  24:1 v/v/v)
- Ribonuclease inhibitor (RNasin): 40 U/µl
- Ribonucleotide (rNTP) mix: 10 mM ATP; 10 mM CTP; 10 mM
  GTP; 1.25 mM UTP. Store at $-20\,°C$
- RNA dyes: 0.4% (w/v) each of bromphenol blue and xylene
  cyanol in 1 mM EDTA, 50% (v/v) glycerol
- Scintillation fluid (T-fluor): 13.5 mM PPO; 0.8 mM POPOP in
  toluol
- Sephadex G-50 spun column (see Sect. 8.4, "Sephadex G-50
  Spun Column")
- T3 RNA polymerase: 15 U/µl

- TBE buffer (1×): 89 mM Tris-OH; 89 mM boric acid; 2 mM EDTA
- TE buffer: 10 mM Tris-HCl, pH 7.9; 1 mM EDTA
- Transcription buffer (5×): 200 mM Tris-HCl, pH 7.5; 30 mM $MgCl_2$; 10 mM spermidine; 50 mM NaCl. Store at $-20\,°C$.

Procedure

Digestion of pBSAL4 with EcoRI

Restriction enzyme digestion

The pBSAL4 plasmid does not contain transcription termination signals. To synthesize RNA of discrete length the plasmid is linearized at a specific site. The RNA polymerase will terminate the synthesis at the 3' end of the template, thus generating a "runoff" transcript. The unique EcoRI restriction site of pBSAL4 (Fig. 8.3) is used for linearization. To determine whether the digest is complete, an aliquot of the sample should be analyzed by agarose gel electrophoresis.

1. Pipet the restriction digestion reaction:
   - 10 µl pBSAL4 (5 µg)
   - 10 µl 2× restriction digestion buffer
   - 2 µl EcoRI (25 U/µl)
   - 28 µl H_2O

2. Incubate for 2 h at 37 °C. Analyze a 5 µl aliquot on a 1 % agarose minigel (in 0.5 × TBE) next to undigested pBSAL4 DNA (see Sect. 8.4, "Preparation of Agarose Minigels").

3. Adjust the volume of the remaining 45 µl of the reaction to 100 µl with H_2O. Add 100 µl of phenol, vortex until an emulsion forms and spin for 1 min at 12 000 g in a microfuge.

4. Transfer the aqueous (upper) phase to a fresh Eppendorf tube and add 100 µl chloroform. Vortex and spin for 1 min at 12 000 g in a microfuge. Transfer the aqueous (upper) phase to a fresh Eppendorf tube. Add 10 µl 3 M sodium acetate and 200 µl 96 % ethanol.

5. Centrifuge the sample for 15 min at 12 000 g and 4 °C in a microfuge, remove the supernatant and dry the pellet.

6. Dissolve the pellet in 90 µl TE buffer (final concentration of 0.5 µg/µl).

In Vitro Transcription

Note: Take the appropriate precautions for working with radioactive material.

1. Combine the following constituents in an Eppendorf tube in the order given below. Keep the tube at room temperature; final volume 30 µl.
   - 5× transcription buffer: 6 µl
   - DTT (0.75 M): 1 µl
   - RNasin (40 U/µl): 1 µl
   - m^{7}GpppG (10 mM): 5 µl
   - rNTP mix: 4 µl
   - NaCl (1 M): 1.5 µl
   - H$_2$O: 7.5 µl
   - [α^{32}P] UTP (7.5×10^5 Bq/µl): 1 µl
   - pBSAL4 DNA (0.5 µg/µl) digested with *EcoRI*: 2 µl
   - T3 RNA polymerase (15 U/µl)

2. Incubate for 1 h at 37 °C.

3. Prepare a Sephadex G-50 spun column (see Sect. 8.4, "Sephadex G-50 Spun Column").

4. Add 1 µl DNase to the reaction and continue the incubation at 37 °C for 10 min.

5. Add 170 µl H$_2$O and 200 µl phenol/chloroform. Vortex for 1 min and spin briefly in a microfuge.

6. Load the aqueous phase slowly (dropwise) onto the G-50 spun column. Cover the column with parafilm and spin at 1600 g for 10 min (Sorvall centrifuge, HB-4 rotor). Collect the flow-through in an Eppendorf tube and estimate the volume.

7. To the flow-through add 1/25 volume 5 M NaCl and 3 volumes 96 % ethanol. Mix well and leave at −80 °C for 1 h.

   Note: If the RNA has a high specific activity, no further purification is necessary. Go directly to step 15. If further purification is required, for example when RNA products of different lengths are synthesized, include steps 8–14.

8. Prepare a denaturing polyacrylamide gel (see Sect. 8.4, "Preparation of Native Polyacrylamide/Agarose Gels").

9. Spin the RNA sample at 4 °C for 10 min, remove the supernatant and dry the pellet. Resuspend the pellet in 10 µl deionized

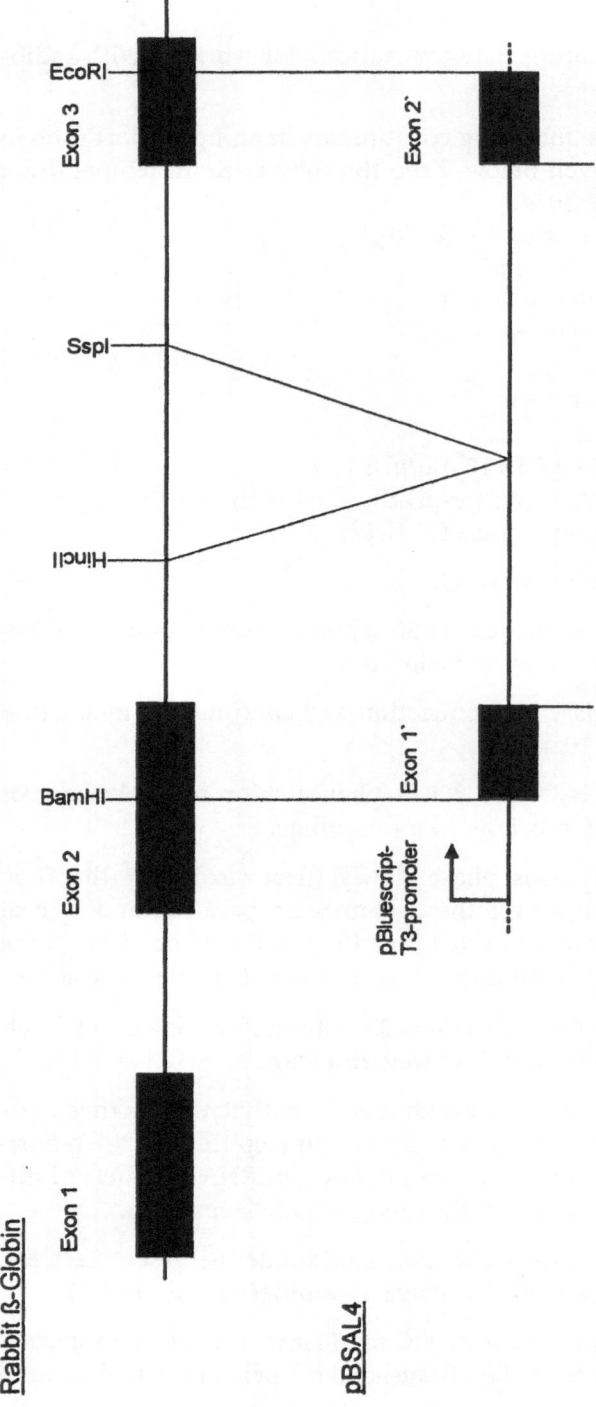

Fig. 8.3. Construction of pBSAL4. A plasmid template for the synthesis of β-globin pre-mRNA was constructed by a three-way ligation between the *Bam*HI-*Hinc*II and *Ssp*I-*Eco*RI fragments of the rabbit β-globin gene and the pBluescript vector which was cleaved in the polylinker at the *Bam*HI and *Eco*RI sites. The resulting construct (pBSAL4) contains wild type 5' and 3' splice sites but has 254 nucleotides of internal intron sequences deleted. The T3 RNA polymerase promoter is indicated upstream of the 5' exon. Globin exon sequences are shown as *boxes*, the intron as a *line*

formamide and add 2 μl RNA dyes. Heat at 100 °C for 2 min. Place the sample on ice immediately.

10. Load the sample onto the RNA gel. Run at 14 W for 2 h, until the xylene cyanol reaches the bottom of the gel.

11. Remove one plate and cover the gel with Saran wrap. Place an X-ray film on the gel and expose for less than 1 min. Mark the position of the film on the Saran wrap with a marker pen (be careful not to move the film). Develop the film.

12. Place the film below the glass plate according to the pen marks. Excise the gel slice that corresponds to the radioactive full-length RNA with a sterile scalpel or razor blade. Place the slice onto a clean glass plate, remove the piece of Saran-wrap and cut the slice into small pieces. Check the radioactivity in the gel slice (should be >2000 cps) and of the remaining gel (should be 2000 cps) with a hand-monitor.

13. Transfer the gel pieces into a fresh Eppendorf tube. Add 300 μl crush-and-soak solution and incubate at 4 °C overnight with constant agitation. Spin briefly to sediment the gel pieces.

14. Load the solution onto a freshly prepared G-50 spun column. Rinse the remaining gel pieces with 100 μl of crush-and-soak solution and load the solution onto the column. Spin the column at 1600 rpm for 10 min and collect the flow-through (should contain more than 2000 cps). Estimate the volume and precipitate the RNA as described in step 7.

15. Spin the RNA sample at 12 000 g in a microfuge, remove the supernatant and dry the pellet. Dissolve the RNA in 20 μl H_2O. Remove 1 μl, dilute 1:10 in H_2O and spot 1 μl of the dilution onto a GF/C glass fiber filter. Measure the Cerenkov counts in 5 ml scintillation fluid and determine the radioactivity in 1 μl of the undiluted RNA sample. For the assay 2×10^5 Cerenkov cpm are needed, which corresponds to 4–40 fmol RNA.

8.3
In Vitro Splicing Reaction

The different stages of the splicing reaction can be analyzed by incubating radiolabeled pre-mRNA in HeLa cell nuclear extracts. The reaction proceeds in the presence of $MgCl_2$ (at an optimal concentration of ~ 3 mM) and KCl (30–60 mM) using ATP as an energy source. Creatine phosphate is added to regenerate ATP from ADP with creatine kinase that is present in the extract. The final volume of the reaction (20 µl) comprises 10 µl of nuclear extract (diluted in buffer D) and 10 µl of a pre-mix which contains the labeled RNA and the remaining substrates (Lamond et al. 1987). The formation of splicing complexes is analyzed in native polyacrylamide gels. Pre-mRNA, intermediates and products of the splicing reaction can be resolved in denaturing polyacrylamide gels. For additional protocols describing in vitro splicing reactions see Krämer and Keller (1990) and Eperon and Krainer (1994). A detailed description of native gel electrophoresis of splicing complexes has been published by Konarska (1989).

In the example given below a timecourse of the splicing reaction (0, 30 and 60 min) will be analyzed. In addition, one reaction will be performed in the absence of ATP to demonstrate the dependence of splicing on energy in the form of ATP. Because the HeLa cell nuclear extract contains endogenous ATP, for this particular experiment the ATP has to be degraded prior to the reaction, which can easily be achieved by pre-incubating the extract for 20 min at 30 °C.

▨ Materials

– Denaturing 7 % polyacrylamide/7 M urea gel (see Sect. 8.4, "Preparation of a Denaturing Polyacrylamide Gel")
– Microfuge
– Native 4 % polyacrylamide/0.5 % agarose gel (see Sect. 8.4, "Preparation of Native Polyacrylamide/Agarose Gels")
– Water bath (30 °C)
– Whatman 3MM paper
– X-ray film and developing equipment

Reagents and
buffers
– Ammonium acetate: 5 M
– ATP: 25 mM

- Buffer D/MgCl$_2$: 20 mM Hepes-KOH, pH 7.9; 20 % (v/v) glycerol; 0.1 M KCl; 0.2 mM EDTA; 6 mM MgCl$_2$; 0.5 mM DTT; 0.5 mM PMSF
- Creatine phosphate: 200 mM
- Ethanol: 96 %
- Deionized formamide
- HeLa cell nuclear extract (see Sect. 8.1, "Preparation of HeLa Cell Nuclear Extracts")
- MgCl$_2$: 50 mM
- ^{32}P-labeled pre-mRNA (see Sect. 8.2, "In Vitro Transcription")
- ^{32}P-labeled pBR322/*Hpa*II DNA marker (see Sect. 8.4, "Preparation of a ^{32}P-Labeled DNA Size Marker")
- Ribonuclease inhibitor (RNasin): 40 U/µl
- RNA dyes: 0.4 % (w/v) each bromphenol blue and xylene cyanol in 1 mM EDTA and 50 % (v/v) glycerol
- Phenol/chloroform: phenol:chloroform:isoamyl alcohol (25:24:1)
- Proteinase K: 20 mg/ml
- Proteinase K buffer: 50 mM Tris-HCl, pH 8.0; 10 mM EDTA; 10 mM NaCl
- SDS: 0.2 %
- Yeast tRNA: 10 mg/ml

Procedure

Note: Take the appropriate precautions for handling radioactive material.   **Splicing reaction**

1. Prepare a native and a denaturing RNA gel.

2. Prepare the pre-mix on ice. Pipet the constituents in the given order. (The pre-mix is calculated for one additional assay to adjust for imprecise pipetting.)

	1×	4×	−ATP (2×)
RNasin (40 U/µl)	0.5 µl	2 µl	1 µl
Creatine phosphate (200 mM)	0.6 µl	2.4 µl	–
ATP (25 mM)	1.45 µl	5.8 µl	–
MgCl$_2$ (50 mM)	1.2 µl	4.8 µl	2.4 µl
pre-mRNA (50 000 cpm/assay)	X µl	4X µl	2X µl
Add H$_2$O to a final volume of	10 µl	40 µl	20 µl

Spin briefly. Pipet 10 µl per assay into labeled Eppendorf tubes. Keep on ice.

3. Mix 30 µl of extract and 30 µl buffer D/MgCl$_2$ in an Eppendorf tube. Spin to clarify for 1 min at 4 °C. Transfer the supernatant to a fresh tube and incubate for 20 min at 30 °C to deplete the extract of ATP.

4. Place the extract on ice and pipet 10 µl aliquots of the extract into the tubes containing the pre-mix. Start the splicing reaction by placing the tubes at 30 °C, in the following order:
   – 60 min (−ATP)
   – 60 min
   – 30 min
   – 0 min (leave on ice)

5. After the incubation spin the reactions briefly at 4 °C and place the tubes on ice.

6. Load a 5 µl aliquot of each reaction onto the native gel. Load 5 µl of RNA dyes into two wells on both sides of the samples. Run the gel at 4 W and 4 °C for about 5.5 h (approximately until 1 h after the xylene cyanol dye has reached the bottom of the gel). Remove the siliconized glass plate. Place two sheets of Whatman 3MM paper onto the gel and lift it up carefully from the glass plate. Dry the gel in a gel dryer and expose it to X-ray film at −80 °C.

7. To the remaining 15 µl of each splicing reaction add 180 µl proteinase K buffer containing 50 µg proteinase K and 10 µg tRNA. Incubate at 30 °C for 10 min.

8. Add 200 µl phenol/chloroform, vortex 1 min and spin briefly. Remove the aqueous phase. Precipitate the RNA by adding 24 µl ammonium acetate and 600 µl 96 % ethanol. Mix and store at −20 °C for 30 min or overnight.

9. Spin for 10 min at 12 000 g and 4 °C in a microfuge, remove the supernatants and dry the pellets. Resuspend in 10 µl deionized formamide and 2 µl RNA dyes. Heat at 100 °C for 2 min and place on ice immediately. Load the samples onto the prepared denaturing polyacrylamide gel. To determine the size of the splicing products load an aliquot of ^{32}P-labeled pBR322/HpaII DNA marker (30 000–50 000 cpm; see "Preparation of α ^{32}P-Labeled DNA Size Marker"). Run the gel at 13 W for 2 h (until the xylene cyanol is about 12 cm from the wells). Remove one

glass plate. Transfer the gel onto a Whatman 3MM paper and dry in a gel dryer. Alternatively, the gel can be covered with Saran wrap on the gel plate. Expose the gel to X-ray film at $-80\,°C$.

8.4
Supplementary Methods

Materials

– 15 ml Corex or screw-cap tubes   **Equipment**
– Glass wool, sterilized
– Plexiglas screen
– Sorvall centrifuge with a swinging bucket rotor
– 2.5 ml syringe
– 20×20 cm glass plates
– Gel apparatus with power supply
– Spacers and comb: 1 mm
– Microwave oven
– Horizontal gel support with spacers and comb
– Horizontal gel apparatus and power supply
– UV-transilluminator
– Microfuge
– Water bath (37 °C)

– TE buffer: 10 mM Tris-HCl, pH 7.9; 1 mM EDTA   **Reagents and**
– STE buffer: 10 mM Tris-HCl, pH 7.9; 1 mM EDTA, 100 mM   **buffers**
  NaCl
– 30 % (w/v) acrylamide (filtered): acrylamide:bisacrylamide
  = 29:1
– Ammonium peroxidisulfate (APS): 10 %
– TBE buffer (1×): 89 mM Tris-OH; 89 mM boric acid; 2 M
  EDTA
– TEMED
– Urea
– 40 % (w/v) acrylamide (filtered): acrylamide:bisacrylamide
  = 80:1
– Low melting agarose
– Siliconization solution
– Tris/glycine buffer (10×): 50 mM Tris-HCl, pH 8.8; 50 mM gly-
  cine

- Agarose
- Ethidium bromide: 10 mg/ml in H_2O
- Agarose minigel (1.5 %)
- $[\alpha^{32}P]dCTP$ (1×10^6 Bq)
- Chloroform:isoamyl alcohol: 24:1 (v/v)
- *E. coli* DNA polymerase I (Klenow fragment)
- dNTP mix: 0.5 mM each of dATP, dGTP and dTTP
- EDTA: 250 mM
- Ethanol: 96 %
- *Hpa*II restriction endonuclease
- Sodium acetate: 3 M, pH 5.2
- 10x Restriction digestion buffer (usually supplied with the enzyme)
- pBR322 plasmid DNA
- Phenol: saturated in 0.1 M Tris-HCl, pH 8.0

Procedure

Sephadex G-50 Spun Column

This method is used to remove unincorporated nucleotides from labeling reactions. The unincorporated nucleotides will be retained by the column whereas the labeled RNA (or DNA) elutes in the exclusion volume. Spun columns are also useful to quickly change the buffer conditions of a biological sample (Sambrook et al. 1989).

Spun column

1. Use an autoclaved Sephadex G-50 suspension, equilibrated in TE buffer.

2. Plug a small piece of sterile glass wool into the outlet of a 2.5 ml syringe (Fig. 8.4).

3. Place the syringe into a 15 ml Corex or screw-cap tube and fill it up to 2.5 ml with G-50 suspension.

4. Cover the tube with parafilm and spin at 1600 g for 10 min in a swinging bucket rotor (Sorvall HB-4 rotor). Continue to add Sephadex until the packed volume is 1.5 ml.

5. Add 0.1 ml of TE and centrifuge at 1600 g for 10 min.

6. Repeat step 5.

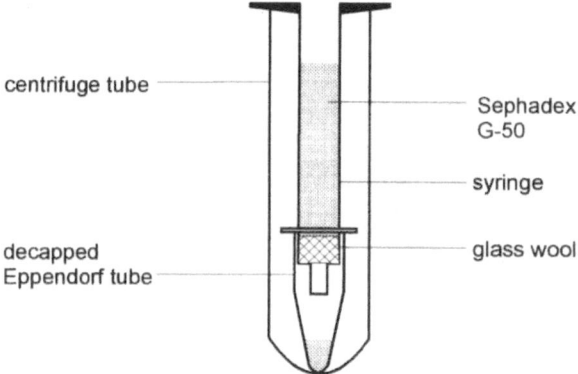

Fig. 8.4. A spun column. The syringe is plugged with a small piece of glass wool and filled with a Sephadex G-50 suspension. An Eppendorf tube without lid is used for the collection of the sample during centrifugation. The Eppendorf tube and the syringe are placed into a 15 ml Corex or screw-cap tube. (Sambrook et al. 1989, reprinted with permission from Cold Spring Harbor Laboratory Press)

7. Discard the flow-through and place an Eppendorf tube without cap at the bottom of the 15 ml tube. Place the outlet of the Sephadex-containing syringe into the Eppendorf tube. The column is now ready to be loaded.

8. Apply the RNA or DNA sample to the column in a total volume of 0.1 ml (use STE to adjust the volume to 0.1 ml).

9. Centrifuge at 1600 g for 10 min, collecting the 0.1 ml of effluent from the syringe in the Eppendorf tube.

10. The unincorporated ^{32}P-dNTPs remain in the syringe, which should be carefully discarded. Labeled RNA or DNA is collected from the Eppendorf tube.

Preparation of a Denaturing Polyacrylamide Gel

1. Assemble a set of 20×20 cm glass plates with 1 mm spacers.

2. Prepare a 7 % acrylamide/7 M urea gel solution:
   - 30 % acrylamide: 8.4 ml
   - Urea: 15.135 g
   - 5× TBE: 7.2 ml
   - Add H_2O to 36 ml.
   - APS: 240 µl
   - TEMED: 24 µl

Denaturing polyacrylamide gel

3. Pour the gel and leave to polymerize for 15–30 min.

4. Pre-run the gel at 14 W for 30 min. Use 0.5× TBE as running buffer.

Preparation of a Native Polyacrylamide/Agarose Gel

Native gel

1. Siliconize one glass plate and assemble plates with 1 mm spacers.

2. Prepare a 4 % acrylamide/0.5 % agarose gel solution as follows:
   - Acrylamide solution: 5 ml 40 % acrylamide; 5 ml 10×Tris-glycine buffer; 20 ml H_2O
   - Agarose solution: 250 mg low melting agarose; 20 ml H_2O

Heat to dissolve agarose (for example in a microwave oven), let cool to ~50 °C. Adjust the volume to 20 ml with H_2O. Thoroughly mix agarose and acrylamide solutions. Add 500 µl APS and 50 µl TEMED.

3. Quickly pour the gel and leave to polymerize overnight at room temperature.

4. Place the gel at 4 °C for about 10 min. Assemble the gel in the gel apparatus and fill the reservoirs with cold 1x Tris-glycine buffer. Carefully remove the comb and wash the wells with buffer. Pre-run the gel at 4 W for 15 min at 4 °C.

Preparation of Agarose Minigels

Agarose minigel

1. Assemble the gel support with spacers and comb.

2. Prepare an agarose gel solution of the desired percentage in 0.5× TBE. Melt the agarose by boiling (for example in a microwave oven). Add water to adjust for the volume lost by evaporation. Add ethidium bromide to a final concentration of 0.5 µg/ml.

3. Pour the gel. Allow the agarose to solidify.

4. Remove spacers and comb.

5. Load the samples. Run the gel at 20 V in 0.5× TBE as running buffer until the bromphenol blue marker has almost reached the end of the gel.

6. Visualize the DNA with a UV-transilluminator.

Preparation of α ^{32}P-Labeled DNA Size Marker

To determine the size of the in vitro splicing products a DNA size marker is electrophoresed next to the samples of the splicing reaction in the denaturing polyacrylamide gel. Because the spliced RNA is visualized by autoradiography, a radiolabeled DNA size marker is used. A commonly used marker is prepared by digestion of pBR322 plasmid DNA with *Hpa*II restriction endonuclease. The fragments generated during the restriction digest are labeled at the 5' ends with the Klenow fragment of *E. coli* DNA polymerase I in the presence of $[\alpha^{32}P]$dCTP.

1. Pipet the restriction digestion mix: **Restriction**
   - pBR322 DNA: 6 µg   **enzyme**
   - *Hpa*II restriction endonuclease: 20 U   **digestion**
   - 10× *Hpa*II restriction digestion buffer: 2 µl
   - Add H$_2$O to 20 µl.

2. Incubate for 2 h at 37 °C.

3. Verify complete digestion by electrophoresis of 3 µl of the restriction digestion mix (corresponding to 1 µg of DNA) in an 1.5 % agarose minigel with 0.5× TBE as running buffer (see Sect. 8.4, "Preparation of Agarose Minigels").

4. Adjust the volume of the remaining 17 µl of the reaction to 100 µl with H$_2$O. Add 100 µl phenol, vortex until an emulsion forms and spin for 1 min at 12 000 g in a microfuge. Transfer the aqueous (upper) phase to a fresh Eppendorf tube and add 100 µl chloroform. Vortex and spin for 1 min at 12 000 g in a microfuge. Transfer the aqueous (upper) phase to a fresh Eppendorf tube. Add 10 µl 3 M sodium acetate and 200 µl 96 % ethanol.

5. Centrifuge the sample for 15 min at 12 000 g and 4 °C in a microfuge, remove the supernatant and dry the pellet.

6. Dissolve the pellet in 20 µl TE buffer.

Note: Take the appropriate precautions for work with radioactive   **End-labeling**
material.   **of DNA**

1. Mix the following:
   - pBR322/*Hpa*II DNA: 3 µl
   - $[\alpha^{32}P]$dCTP (1×10^6 Bq): 3 µl
   - dNTP mix: 2 µl

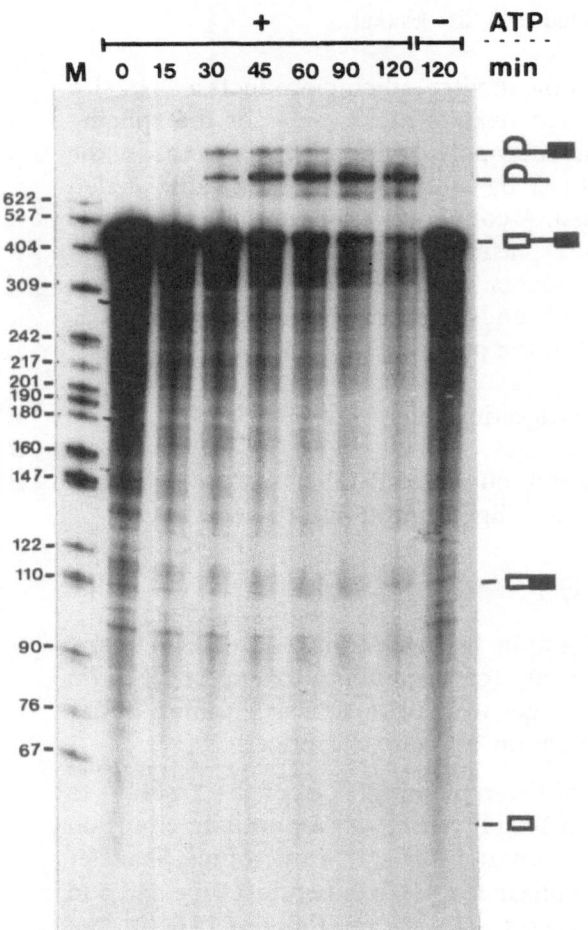

Fig. 8.5. Kinetics of in vitro pre-mRNA splicing. Electrophoretic separation of splicing products in a 7% polyacrylamide/7 M urea gel. Radiolabeled rabbit β-globin pre-mRNA was incubated in a HeLa cell nuclear extract for 0–120 min at 30 °C in the presence of ATP. A control reaction was incubated in the absence of ATP for 120 min. The positions of the pre-mRNA, intermediates and products of the splicing reaction are shown schematically. *M* ^{32}P-labeled pBR322/*Hpa*II DNA size marker. *Open lines* indicate the 5' exon, *closed lines* indicate the 3' exon

- DNA polymerase I (Klenow fragment): 8 U
- Add H_2O to 10 µl.

2. Incubate for 20–25 min at room temperature.

3. Stop the reaction by addition of 2 µl 250 mM EDTA.

4. Add 10 µl TE and store at −20 °C.

5. Count 1 µl of the solution in a scintillation counter. Use 30 000–50 000 Cerenkov cpm per lane for the denaturing gel.

Note: The sizes of the DNA fragments are 622, 527, 404, 309, 242, 238, 217, 201, 190, 180, 160, 147, 122, 110, 90, 76, 67, 34, 26, 15 and 9 base pairs.

Results

- In vitro splicing reaction
  The results of a typical splicing reaction are shown in Fig. 8.5. In the reactions performed for 0 and 15 min, and in the absence of ATP, only pre-mRNA (and degradation products) is visible. After 30 min a product of ~50 nucleotides that corresponds in size to the cleaved-off 5' exon appears. In addition, two products that migrate slower than the pre-mRNA are detected. These products correspond to the intron-3' exon and the excised intron. Because of their branched circular (or lariat) conformation these products are retarded in their migration in the denaturing gel (see also below). At later time points the intensity of the products of the reaction (released intron-lariat and spliced mRNA) increases, whereas a decrease in the intensity of the intermediates is apparent, indicating the conversion of intermediates into products at later times of the splicing reaction.

General Comments. The results of an in vitro splicing reaction may be difficult to interpret, e.g., when a RNA substrate is used for the first time or when a pre-mRNA contains more than one intron. The appearance of different bands in the gel during a time course of a splicing reaction and the electrophoretic mobility of individual bands provide the main guidelines in interpreting the results. Intermediates should appear first and their intensity might decline when the products of the reaction appear. Often the products containing the 3' end of the substrate are subject to nuclease degradation which becomes apparent during the time course. Lariat intron-containing species can be identified by a different migration in gels of different polyacrylamide concentrations. When gels of low percentage polyacrylamide are used, lariat products migrate at approximately the expected sizes. In gels of high percentage polyacrylamide these products are retarded in the gel and often migrate above the pre-mRNA. Moreover, to identify lariat-containing RNAs, the products of a splicing reaction can be isolated from a denaturing gel and treated with a debranching enzyme that specifically cleaves the 2'5'-phosphodiester bond at the lariat branch site thus generating a linear RNA that will migrate according to its expected size even in gels of high percentage polyacrylamide (Ruskin and Green 1985; Grabowski 1994).

- In vitro spliceosome assembly
  Figure 8.6 shows a representative example of the analysis of splicing complexes by native gel electrophoresis. Without incubation at 30 °C (or in the absence of ATP, not shown) only complex H is visible. As described in the Introduction, this complex forms with any RNA substrate, i.e., in the absence of introns, and is considered to be nonspecific. Pre-splicing com-

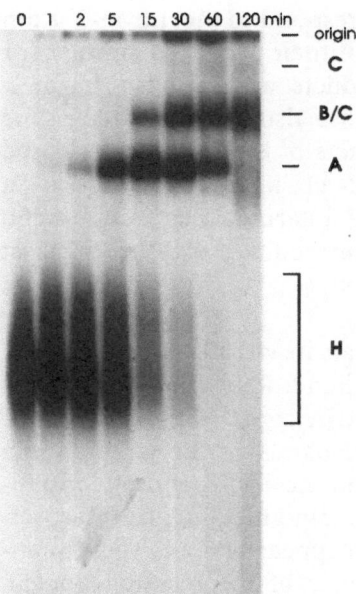

Fig. 8.6. Kinetics of in vitro pre-mRNA splicing. Electrophoretic separation of splicing complexes in a native polyacrylamide gel. Splicing reactions were performed for 0–120 min at 30 °C in the presence of ATP. The positions of complexes H, A, B and C are indicated

plex A forms after 1-2 min of incubation at 30 °C in the presence of ATP and is present at higher concentrations after 5 min. Splicing complex B is detected after 15 min and has accumulated to high levels in the reaction incubated for 120 min. Complex C is visible as a light smear between complex B and the origin of the gel, however, most of this complex is unstable and migrates together with complex B. Consistent with the conversion of pre-splicing complex A into splicing complexes B and C, complex A is reduced in intensity in the reactions performed for 60 and 120 min. In addition to the products detected in the gel shown in Fig. 8.6, the released intron may be visible as a faint band that migrates between complexes A and B/C at late times of the reaction (Konarska and Sharp 1987). The released mRNA migrates in the range of complex H (Frendewey et al. 1987).

Comments

The assignment of different complexes produced during the in vitro splicing reaction can also best be made by following a time course of the reaction, as shown in Fig. 8.6. Control reactions can include one sample incubated in the absence of extract. In such an experiment the pre-mRNA will be visible as a discrete band at the bottom of the gel. In a control reaction performed in the absence of ATP, only complex H will be visible. To analyze the RNA species that are associated with the different splicing complexes, a lane of the native gel can be excised and the RNA species within the native gel slice can then be separated in a second dimension denaturing gel (Frendewey et al. 1987).

References

Beggs JD (1995) Yeast splicing factors and genetic strategies for their analysis. In: Lamond AI (ed) Pre-mRNA processing. RG Landes, Austin, pp 79-95

Chabot B (1994) Synthesis and purification of RNA substrates. In: Higgins SJ, Hames BD (eds) RNA processing: a practical approach, vol I. IRL, Oxford, pp 1-29

Dignam JD, Lebovitz RM, Roeder RG (1983) Accurate transcription initiation by RNA polymerase II in a soluble extract from isolated mammalian nuclei. Nucleic Acids Res 11:1475-1489

Dignam JD (1990) Preparation of extracts from higher eukaryotes. Methods Enzymol 182:194-203

Eperon IC, Krainer AR (1994) Splicing of mRNA precursors in mammalian cells. In: Higgins SJ, Hames BD (eds) RNA processing: a practical approach, vol I. IRL, Oxford, pp 57–101

Frendewey D, Krämer A, Keller W (1987) Different small nuclear ribonucleoprotein particles are involved in different steps of splicing complex formation. CSH Symp. Quant. Biol. 52, 287–298

Grabowski PJ (1994) Characterization of RNA. In: Higgins SJ, Hames BD (eds) RNA processing: a practical approach, vol I. IRL, Oxford, pp 31–55

Green MR, Maniatis T, Melton DA (1983) Human β-globin pre-mRNA synthesized in vitro is accurately spliced in Xenopus oocyte nuclei. Cell 32:681–694

Konarska MM (1989) Analysis of splicing complexes and small nuclear ribonucleoprotein particles by native gel electrophoresis. Methods Enzymol 180:442–453

Konarska MM, Sharp PA (1987) Interactions between small nuclear ribonucleoprotein particles in the formation of spliceosomes. Cell 49:763–774

Krainer AR, Conway GC, Kozak D (1990) Purification and characterization of pre-mRNA splicing factor SF2 from HeLa cells. Genes Dev 4:1158–1171

Krämer A (1996) The structure and function of proteins involved in nuclear pre-mRNA splicing. Annu Rev Biochem 65:367–409

Krämer A, Keller W (1990) Preparation and fractionation of mammalian extracts active in pre-mRNA splicing. Methods Enzymol 181:3–19

Lamond AI, Konarska MM, Sharp PA (1987) A mutational analysis of spliceosome assembly: evidence for splice site collaboration during spliceosome formation. Genes Dev 1:532–543

Lee KA, Green MR (1990) Small-scale preparation of extracts from radiolabeled cells efficient in pre-mRNA splicing. Methods Enzymol 181:20–30

Melton DA, Krieg PA, Rebagliati MR, Maniatis T, Zinn K, Green MR (1984) Efficient in vitro synthesis of biologically active RNA and RNA hybridization probes from plasmids containing a bacteriophage SP6 promoter. Nucleic Acids Res 12:7035–7056

Moore MJ, Query CC, Sharp PA (1993) Splicing of precursors to mRNA by the spliceosome. In: Gesteland RF, Atkins JF (eds) The RNA world. Cold Spring Harbor Laboratory, Cold Spring Harbor, pp 303–357

Newman A (1994) Analysis of pre-mRNA splicing in yeast. In: Higgins SJ, Hames BD (eds) RNA processing: a practical approach, vol I. IRL, Oxford, pp 57–101

Ruskin B, Green MR (1985) An RNA processing activity that debranches RNA lariats. Science 229:135–140

Sambrook J, Fritsch EF, Maniatis T (1989) Molecular cloning: a laboratory manual. Cold Spring Harbor Laboratory, Cold Spring Harbor

Yisraeli JK, Melton DA (1989) Synthesis of long, capped transcripts in vitro by SP6 and T7 RNA polymerases. Methods Enzymol 180:42–50

Zahler AM, Lane WS, Stolk JA, Roth MB (1992) SR proteins: a conserved family of pre-mRNA splicing factors. Genes Dev 6:837–847

Subject Index

Abbreviations

(without SI-units, symbols in equations or chemical elements)

A	absorption, Adenosine, U1 snRNP protein A	CENP	Centromer Associated Proteins
A'	U1 snRNP protein A'	CIE	Counter Immuno Electrophoresis
A_1	hnRNP core protein A_1	CM	Carbon-coated Mica
A_2	hnRNP core protein A_2	COMET	Constrained Maximum Entropy Tomography
ANA	anti-nuclear Antibody	CP	Creatine Phosphate
AP	Alkaline Phosphatase	CTE	Calf Thymus Extract
APS	Ammonium Persulphate	CTP	Cytidine 5'-Triphosphate
ATCC	American Type Cell Culture	1-D	one-dimensional
		2-D	two-dimensional
ATP	Adenosine 5'-Triphosphate	3-D	three-dimensional
		D1	common snRNP protein D1
B	common snRNP protein B	D2	common snRNP protein D2
B'	common snRNP protein B'	D3	common snRNP protein D3
B"	U2 snRNP protein B"		
B_1	hnRNP core protein B_1	d	deoxy
B_2	hnRNP core protein B_2	dcp	Density Correlation Program(s)
BCIP	5-Bromo-4-chloro-3-indoxyl Phosphate	DE53	Resin DE53
BR	Balbiani Ring	DEC	Digital Equipment Corporation
BS	Bovine Serum	DEPC	Diethyl Pyrocarbonate
BSA	Bovine Serum Albumine	DHFR	Dihydrofolate Reductase
C	Cytosine, U1 snRNP protein C	DMF	N,N-Dimethyl-formamide
C_1	hnRNP core protein C_1	DMSO	Dimethyl Sulfoxide
C_2	hnRNP core protein C_2	DNA	Deoxyribonucleic Acid
CAD	Carbamoyl-P-Synthetase, Aspartate Transcarbamylase, Dihyrdo-Orotase	DSP	5,3'-Dithio-bis(Propionic Acid N-Hydroxysuccin-imide Ester)
		DTE	Dithioerythirol
cDNA	complementary DNA	DTT	Dithiothreitol

E	energy, common snRNP protein E	Mab	monoclonal antibody
EDTA	Ethylenediamine tetraacetic Acid	MCTD	Mixed Connective Tissue Disease
ELISA	Enzyme Linked Immunosorbent Assay	β-ME	β-Mercaptoethanol
		min	minute(s)
EM	Electron Microscope, Electron Microscopy	Mono Q	Mono Q Chromatography
Eq	equation	mRNA	messenger Ribonucleic Acid
ET	Electron Tomography	N	(any) Nucleotide
F	fluorescence, common snRNP protein F	NBT	4-Nitroblue Tetracolium Chloride
FITC	Fluoresceine Isothiocyanate	NEpHGE	Nonequilibrium pH Gradient Gel Electrophoresis
FPLC	Fast Protein Liquid Chromatography	NET-2	see buffers
G	Guanosine, common snRNP protein G	NFS	Network File System
		NHS	see buffers
G-50	Sephadex G-50	NP-40	Nonidet P-40
GSB	see buffers	OASE	Oligoadenylate Synthetase
GTC	Guanidinium Thiocyanate	OD_{260}	Optical Density (at 260 nm wavelength)
GTP	Guanosine 5'-Triphosphate		
H	Histone	PAA	Polyacrylamide
h	hour(s)	PAGE	Polyacrylamide Gel Electrophoresis
HEPES	N-(2-Hydroxyethyl)piperazine-N'-(2-ethanesulfonic acid)	PAS	Protein A-Sepharose
		PAS-AB	Protein A-Sepharose-Antibody-Complex
HIV	Human Immunodeficiency Virus	PAS-AB-AG	Protein A-Sepharose-Antibody-Antigen-Complex
hnRNA	heterogeneous nuclear Ribonucleic Acid		
hnRNP	heterogeneous nuclear Ribonucleoprotein (Particle)	PBC	Primary Billiary Cirosis
		PBS	see buffers
HSE	Human Spleen Extract	PC	personal computer
I	intensity	PCR	Polymerase Chain Reaction
Ig	Immunoglobuline		
IEF	Isoelectric Focussing	PCV	Packed Cell Volume
IEP	Isoelectric Point	PEG	Polyethylene Glycol
K	binding constant	Phe	Phenylalanine
K_{sv}	Stern-Vollmer constant	PIPES	Piperazine-N,N'-bis(2-ethanesulfonic acid)
Kac	Potassium Acetate		
L	loose fitting (homogenizer)	PMSF	Phenylmethylsolfonyl Fluoride
La	patients autoimmune serum La	PM	Polymyosis
		pNPP	p-Nitrophenyl Phosphate
lnRNP	large nuclear Ribonucleoprotein (Particle)	PNV	Packed nuclear Volume
m_3G-cap	Trimethylguanosine-Cap		

POCS	Projection onto convex Sets	snRNA	small nuclear Ribonucleic Acid
poly A⁺	polyadenylated	snRNP	small nuclear Ribonucleoprotein (Particle)
POPOP	1,4-bis(5-Phenyl-2-oxazolyl)-benzene		
PPO	2,5-Diphenyloxazole	SONB	see buffers
pre-mRNA	precursor of messenger Ribonucleic Acid	SR	Serine/Arginine rich Protein
pre-mRNP	pre-messenger Ribonucleoprotein (Particle)	SS	Sjorgens Syndrome
		SSA	patients autoimmune serum SSA
PSF	Polypyrimidine Tract-binding Protein-associated Splicing Factor	ST2M	see buffers
		STE	see buffers
		t	time
		TBE	see buffers
PTB	Polypyrimidine binding Protein	TBS	see buffers
		TCA	Trichloroacetic Acid
Py	Pyridine	TE	see buffers
Q	Quencher	TEMED	N,N,N',N'-Tetramethyenediamine
R	Purine		
RA	Rheumatoid Arthritis	TMV	Tobacco Mosaic Virus
RIA	Radio Immuno Assay	Tris	Tris(hydroxymethyl) aminomethane
Ro	patients autoimmune serum Ro		
		Tris-HCl	Tris buffer, pH adjusted with HCl
RNA	Ribonucleic Acid		
rRNA	ribosomal RNA	tRNA	transfer Ribonucleic Acid
RNase	Ribonuclease		
RNasin	RNase Inhibitor	Trp	Tryptophane
RNP	Ribonucleoprotein (Particle)	TSS	see buffers
		Tyr	Tyrosine
rNTP	Ribonucleotide Triphosphate	U	Uridine
		U1	U1 snRNA
RT-PCR	Reverse Transcriptase Polymerase Chain Reaction	U2	U2 snRNA
		U2AF	U2 snRNP auxiliary factor
S	tight fitting (homogenizer)	U3	U3 snRNA
		U3-snoRNP	U3 small nucleolar RNP
s	second(s)		
SB	see buffers	U4	U4 snRNA
Scl	Scleroderma	U5	U5 snRNA
Scl 70	Topoisomerase I	U6	U6 snRNA
SDS	Lauryl Sulfate, Sodium Salt	UTP	Uridine 5'-Triphosphate
SDS-PAGE	SDS-Polyacrylamide Gel Electrophpresis	UV	ultra violet
		VR	Vanadyl Ribonucleoside Complex
SF	Splicing Factor		
SLE	Systemic Lupus Erythematosus	WB	see buffers
		Y	Pyrimidine
Sm	patients autoimmune serum Sm	z	density in the beam